The Foundations of Electric Circuit Theory

The Foundations of Electric Circuit Theory

N R Sree Harsha
South West Nuclear Hub and Bristol–Oxford Nuclear Research Centre, University of Bristol, UK

Anupama Prakash
Amity School of Engineering, Uttar Pradesh, India

D P Kothari
Former Director I/C, IIT Delhi,
former Vice Chancellor, VIT Vellore,
former Principal, VRCE Nagpur

IOP Publishing, Bristol, UK

ISBN 978-0-7503-1266-0 (ebook)
ISBN 978-0-7503-1267-7 (print)
ISBN 978-0-7503-1268-4 (mobi)

DOI 10.1088/978-0-7503-1266-0

Version: 20161001

IOP Expanding Physics
ISSN 2053-2563 (online)
ISSN 2054-7315 (print)

British Library Cataloguing-in-Publication Data: A catalogue record for this book is available from the British Library.

Published by IOP Publishing, wholly owned by The Institute of Physics, London

IOP Publishing, Temple Circus, Temple Way, Bristol, BS1 6HG, UK

US Office: IOP Publishing, Inc., 190 North Independence Mall West, Suite 601, Philadelphia, PA 19106, USA

To my grandmother, V Ramalakshmi. I would much prefer it if you were alive.
N R Sree Harsha
To my mother, Smt. Usha Gupta. Anupama Prakash
To my daughters Seema and Shikha. D P Kothari

'It is fascinating to muse: Would Faraday have discovered the law of electromagnetic induction if he had received a regular college education? Unencumbered by the traditional way of thinking, he felt that the introduction of the "field" as an independent element of reality helped him to coordinate the experimental facts.'

Albert Einstein 1950 On the generalized theory of Gravitation *Sci. Am.* **182** 4

Contents

Preface

The traditional textbooks on electricity and magnetism treat 'classical electromagnetism' and 'electrical circuit analysis' as separate and disjoint subjects. This way of presentation can lead to a wrong perception that these are two separate topics that have different sets of fundamentals. This book, we believe, bridges this gap in the usual presentation of these two fundamentally important subjects and gives the reader a deeper understanding of how electric circuits behave from the standpoint of the fundamental concepts of the classical electromagnetism.

In classical electromagnetism, electrostatic phenomena are generally analyzed using fields and charges. In electric circuit theories, currents and potentials are used to analyze electric circuits. By providing a smoother transition to electric circuit analysis from electromagnetism, we can better understand the fundamental theories in electric circuit analysis.

Chapter 1 presents the necessary mathematical tools that are needed to better understand the physical concepts presented in the book. It is divided into two sections.

The first section presents the calculus of variations and the second section presents the vector algebra and vector calculus.

Chapter 2 introduces the concept of *charge* and different processes of electrification. The triboelectric series is presented to determine the charge acquired by different materials by contact electrification. Coulomb's law is then presented which becomes the foundation for our further understanding of the electric phenomena.

In chapter 3 we shall see how adding velocity and acceleration terms to Coulomb's law does not give us a simple and concise view of the interaction of charges. The laws of electromagnetism can be expressed in the simplest form by introduction of 'fields'. This chapter concludes with integral and differential form of Gauss' law.

Chapter 4 introduces the concept of electric potential and potential energy of a system of charges. Earnshaw's theorem is presented and we prove that 'it is not possible to have a stable array of charged particles in static, stable equilibrium under the influence of electric and magnetic forces alone'. The concept of a capacitor is introduced and its behavior in various situations is studied.

Chapter 5 introduces the special theory of relativity and the kinematic consequences of its postulates that are relevant to electric circuits are presented. Ohm's law is presented and two types of the surface charges are discussed. Surface charges are present at the boundary between a current carrying conductor and the medium in which it is present (usually air), and provide the necessary electric fields inside the conductors for the current to flow. The chapter concludes with the application of the Poynting vector in the case of a simple DC circuit and, a rather surprising result, that ideal conductors in a simple DC circuit carry no energy from the battery to the resistor.

Chapter 6 introduces the two methods to calculate the magnetic field produced by the current carrying conductors. The Lorentz contraction is used to show why (not how!) a magnetic field is created around a current carrying conductor. The chapter concludes with Rowland and Hall's experiments and we will learn why we know that only electrons constitute the electric current in the metallic conductors.

Chapter 7 introduces Faraday's experiments and the phenomenon of electromagnetic induction. The concepts of self and mutual induction and the properties of an inductor are presented.

Chapter 8 introduces two problems that result by applying Ampère's circuital law. A new term called 'displacement current' is introduced to increase the scope of Ampère's law. The full set of Maxwell's equations is presented and Helmholtz's theorem is stated. The choice of gauge in electromagnetic theory is presented and, as we shall see, this answers important questions such as 'does displacement current really create a magnetic field?'

Chapter 9 introduces Kirchhoff's laws (KVL and KCL) and the derivation of Kirchhoff's laws for the case of a simple circuit is presented. The concept of *entropy* in an electric circuit is presented and Maximum Entropy Production Principle is used to solve DC circuits without applying KVL. A number of different circuit analysis tools are then presented.

Acknowledgements

We would like to thank everybody who helped and assisted during the writing. In particular, we thank John Navas of IOP who reposed his trust in us. N R Sree Harsha would like to thank his family and friends for their love and support. Anupama Prakash would like to thank her husband, Sandeep, and her daughter, Shreya, for their unstinting support and encouragement. We thank the editors and designers at IOP who assisted in editing, proofreading and design.

Author biographies

N R Sree Harsha

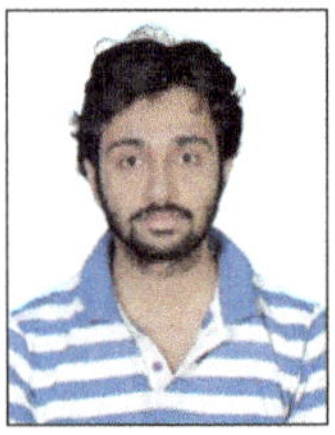

N R Sree Harsha obtained his undergraduate degree in Electrical and Electronics Engineering from R V College of Engineering, Bangalore, India. He is currently pursuing a Masters in Nuclear Science and Engineering at the University of Bristol, UK. He loves to sit in a café alone and read a Stephen King or Richard Dawkins book. In his spare time, he loves to tell people how beautiful Nature is.

Anupama Prakash

Anupama Prakash is an Associate Professor in the Department of Electrical & Electronics Engineering, Amity School of Engineering, Amity University, Noida (U.P.). She has 18 years of teaching experience in the field of electrical engineering. She received a Bachelors of Engineering from Madan Mohan Malviya Engineering College in 1991, and went on to complete her Masters in Technology from Jamia Millia Islamia, New Delhi. In 2012, she was awarded a PhD from Jamia Millia Islamia, New Delhi.

She has done extensive research work in the areas of Power Systems and Electric Circuits. Her research work has been published in various international journals and reputed Conferences.

Dr D P Kothari

Dr D P Kothari obtained his BE(Electrical) in 1967, ME(Power Systems) in 1969 and PhD in 1975 from Birla Institue of Technology and Sciences (BITS), Pilani, Rajasthan. From 1969 to 1977, he was involved in teaching and development of several courses at BITS Pilani. Prior to assuming charge as Director Research of GPGI, Nagpur, Dr Kothari served as Vice Chancellor, VIT, Vellore, Director in-charge and Deputy Director (Administration) as well as Head in the Centre of Energy Studies at Indian Institute of Technology, Delhi and as Principal, Visvesvaraya Regional College of Engineering, Nagpur. He was visiting professor at the Royal Melbourne Institute of Technology, Melbourne, Australia, during 1982–83 and 1989, for two years. He was also NSF Fellow at Perdue University, USA in 1992.

Dr Kothari, who is a recipient of the most Active Researcher Award, has published and presented 780 research papers in various national as well as international journals, conferences, guided 48 PhD scholars and 68 M. Tech students, and authored 45 books in various allied areas. He has delivered several keynote addresses and invited lectures at both national and international conferences. He has also delivered 42 video lectures on YouTube with a maximum of 40 000 hits!

Dr Kothari is a Fellow of the National Academy of Engineering (FNAE), Fellow of Indian National Academy of Science (FNASc), Fellow of Institution of Engineers (FIE), Fellow IEEE and Hon. Fellow ISTE .

His many awards include the National Khosla Award for Lifetime Achievements in Engineering (2005) from IIT, Roorkee. The University Grants Commission (UGC), Government of India has bestowed the UGC National Swami Pranavandana Saraswati Award (2005) in the field of education for his outstanding scholarly contributions.

He is also the recipient of the Lifetime Achievement Award (2009) conferred by the World Management Congress, New Delhi, for his contribution to the areas of educational planning and administration. Recently he received Excellent Academic Award at IIT Guwahati by NPSC-2014.

IOP Publishing

The Foundations of Electric Circuit Theory

N R Sree Harsha, Anupama Prakash and D P Kothari

Chapter 1

Mathematical introduction

1.1 Introduction to the calculus of variations

Consider two quantities y and x that are related in such a way that for every value of x, there is one and only one value of y. The different values of x for which y is well defined and single valued are called the domain and this is often represented as D. We call such a relationship a 'functional relationship' and we represent it as $y = f(x)$. In dealing with the function of a single variable such as $y = f(x)$, it is often desirable to determine the values of x for which the function y attains a local maximum or minimum value. This is predominantly what the calculus of variations is about. This field is as old as calculus and has attracted the attention of some of the greatest mathematical scientists, such as Newton, Leibniz, Lagrange, the Bernoulli brothers, Jacobi, etc. Its applications are very wide and it continues to provide solutions to many new physical and engineering problems. We shall also be using the calculus of variations later in our book to apply the maximum entropy production principle (MEPP) in DC circuits. In this section, we shall first study the calculus of a single variable and later move on to many variables.

1.1.1 Absolute extremum

If the function $f(x)$ achieves a local maximum at x_1, this means that the value of the function f in the immediate neighborhood of x_1 is less than or equal to $f(x_1)$. A similar definition also holds for a local minimum. The maximum and minimum points of the function are generally termed as the extremum points of the function. If, on the other hand, the function f achieves an extremum value at a point x_0 for all the values of $f(x)$, with x belonging to the D, we say that the extremum value is an 'absolute' or 'global' extremum. Hence, we have the following definition:

Suppose $y = f(x)$ is defined in the domain D. We say that a function $f(x)$ attains its absolute maximum (or minimum) value at the point x_0 if $f(x) \leqslant f(x_0)$ (or $f(x) \geqslant f(x_0)$) for all $x \in D$.

doi:10.1088/978-0-7503-1266-0ch1

Let us suppose, in order to understand clearly, that the function $y = f(x)$ is as shown in figure 1.1. The function f achieves global maximum at x_1 and global minimum at x_2. It is clear from figure 1.1 that the tangent to the curve at these extrema points is parallel to the x-axis. This implies that the slope of the curve is zero. However, the slope of the curve at a point x is given by $\mathrm{d}y/\mathrm{d}x$. Hence, at these extrema points, we have that

$$\left.\frac{\mathrm{d}y}{\mathrm{d}x}\right|_{x_1} = \left.\frac{\mathrm{d}y}{\mathrm{d}x}\right|_{x_2} = 0.$$

It is also clear from figure 1.1 that just before x_1 the slope of the function is positive, while the slope is zero at x_1, and it is negative just after x_1. Hence, we see that the value of the slope of $f(x)$ decreases while passing from A to C in figure 1.1. This implies that

$$\left.\frac{\mathrm{d}}{\mathrm{d}x}\left[\text{slope}\right]\right|_{x_1} = \left.\frac{\mathrm{d}}{\mathrm{d}x}\left[\frac{\mathrm{d}y}{\mathrm{d}x}\right]\right|_{x_1} = \left.\frac{\mathrm{d}^2y}{\partial x^2}\right|_{x_1} < 0.$$

After applying a similar line of thought for the global minimum, we see that

$$\left.\frac{\mathrm{d}^2y}{\partial x^2}\right|_{x_2} > 0.$$

These are sufficient and necessary conditions to determine the global extrema points of a function.

We shall now consider a function of many variables. If x_0 is an extremum of a function $f(x_1, x_2, x_3, \ldots x_n)$, then at this point each partial derivative is either equal to zero or does not exist. Thus, at x_0 we have that

$$\left.\left[\frac{\partial f}{\partial x_1} = \frac{\partial f}{\partial x_2} = \frac{\partial f}{\partial x_3} \cdots \frac{\partial f}{\partial x_n}\right]\right|_{x_0} = 0.$$

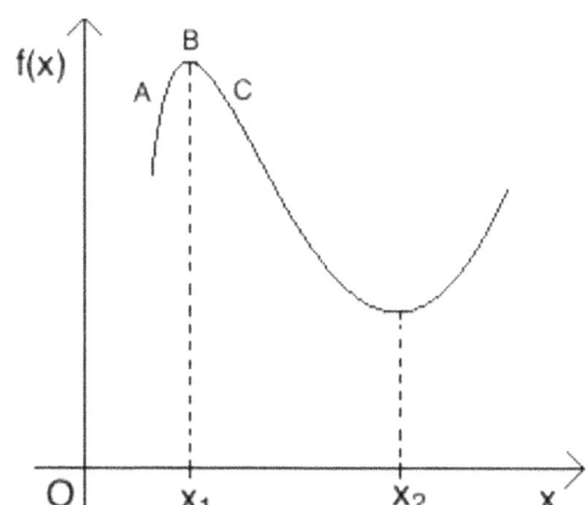

Figure 1.1. A graph showing the relationship of function $f(x)$ with its independent variable x. The function $f(x)$ attains global maximum at point x_1 and global minimum at point x_2.

Example 1 Consider the function $f(x, y) = x^2 + y^2$ defined for all $x, y \in Z$ (integers). Find the global extrema point of the function.

Solution

For global extrema points, we have that

$$\frac{\partial f(x, y)}{\partial x} = \frac{\partial f(x, y)}{\partial y} = 0.$$

Next, $\frac{\partial f(x, y)}{\partial x} = 0$ implies that $2x = 0$. This implies that $x = 0$. Similarly, for $\frac{\partial f(x, y)}{\partial y} = 0$, we obtain that $y = 0$. Hence, the point of extremum is at (0, 0).

We see that the origin is a point of global minimum because at (0, 0) we have that

$$\frac{\partial^2 f(x, y)}{\partial x^2} = \frac{\partial^2 f(x, y)}{\partial y^2} = 2 > 0.$$

Example 2 Consider the parallel combination of resistors shown in figure 1.2. A fixed current I enters the combination at node A. Let us assume that the current through the resistor R_1 is i_1. Let us also assume that the current through the resistor R_2 is $(I - i_1)$. Find out the value of i_1 in terms of I, R_1 and R_2 by extremizing the following function:

$$P = i_1^2 R_1 + (I - i_1)^2 R_2.$$

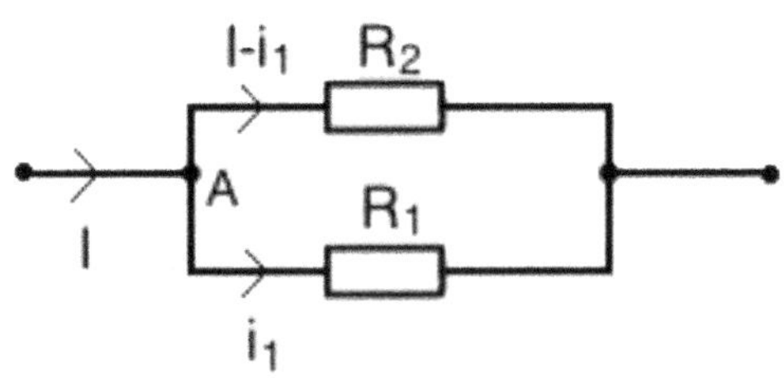

Figure 1.2. A parallel combination of resistors to demonstrate the principle of the least dissipation of power in resistors. The true current distribution is one that minimizes the total ohmic dissipation in resistors.

Solution

The function given is $P(i_1)$ and we have to extremize P with respect to i_1. Hence, we require that

$$\frac{dP}{di_1} = 0$$
$$\Rightarrow 2i_1R_1 - 2(I - i_1)R_2 = 0$$
$$\Rightarrow i_1 = \frac{IR_2}{R_1 + R_2}.$$

Hence, we have that the current that flows through resistor R_1 is given by $i_1 = \frac{IR_2}{R_1 + R_2}$ and the current that flows through resistor R_2 is given by $I - i_1 = \frac{IR_1}{R_1 + R_2}$. For those of you familiar with circuit analysis, you can see that extremizing the power dissipated in the resistors gives us the correct distribution of currents in the parallel combination. Kirchhoff first discovered this in 1848 in a three-dimensional case [1], and later Maxwell and Feynman applied it to the case of electric circuits, as demonstrated in this example [2, 3].

1.1.2 Conditional extremum

Let us now learn how to extremize a function $f(x)$ of many variables subject to certain predefined constraints. Suppose, in order to be more specific, that $y = f(x) = f(x_1, x_2, x_3, \dots x_n)$ is a function of many variables $x_1, x_2, x_3, \dots x_n$ defined in the domain D and that these independent variables are constrained by the following m conditions:

$$\left.\begin{aligned} \varphi_1(x_1, x_2, x_3 \dots x_n) &= 0 \\ \varphi_2(x_1, x_2, x_3 \dots x_n) &= 0 \\ \varphi_3(x_1, x_2, x_3 \dots x_n) &= 0 \\ &\vdots \\ \varphi_m(x_1, x_2, x_3, \dots x_n) &= 0 \end{aligned}\right\} \text{ equations of constraints.}$$

These m equations of the independent variables are called the 'equations of constraints'. Let us also suppose that there is a point $P(x_1^S, x_2^S, x_3^S \dots x_n^S)$ in the domain D such that

$$f(P) \geqslant f(x) \qquad \text{or} \qquad f(P) \leqslant f(x),$$

while simultaneously satisfying the constraints defined by φ_S. Then we say that the function f has a conditional extremum at point P. A common way of finding such points of extrema is called the *Lagrangian method of multipliers*. We shall now see how to use this method to find out the extrema points.

Suppose that the functions $f(x_1, x_2, x_3, \ldots x_n)$ and m equations of constraints $\varphi_{1,2,3\ldots m}(x_1, x_2, x_3 \ldots x_n)$ have continuous first-order partial derivatives in the domain D with $m < n$. A new function called the *Lagrangian function* Ψ is formed such that

$$\Psi = f(x_1, x_2, x_3 \ldots x_n) + \sum_{i=1}^{m} \lambda_i \varphi_i(x_1, x_2, x_3 \ldots x_n),$$

where the λ_is are undetermined constants. Now, the function Ψ is investigated for an absolute extremum by solving the following equations:

$$\frac{\partial \Psi}{\partial x_i} = 0, \tag{1.1}$$

where i in x_i runs from 1 to n. From equations (1.1) and the m equations of constraints, the undetermined λ_is and the coordinates of the possible extrema points are discovered. If a point P is the absolute extremum of the Lagrangian function Ψ, then it is the conditional extremum of the function f. Let us illustrate this with a simple example.

Example 3 Extremize the function $f = (x_2 + y^2)/2$ subject to the condition that $x + y = 1$.

Solution
We have

$$f(x, y) = \frac{x^2 + y^2}{2}$$
$$\varphi(x, y) : x + y - 1 = 0.$$

The Lagrange function is then given by

$$\Psi = \frac{x^2 + y^2}{2} + \lambda(x + y - 1).$$

For the absolute extremum of the Lagrange function, we need to have that

$$\frac{\partial \Psi}{\partial x} = \frac{\partial \Psi}{\partial y} = 0.$$

For $\frac{\partial \Psi}{\partial x} = 0$, we have that $x = -\lambda$ and for $\frac{\partial \Psi}{\partial y} = 0$ we have that $y = -\lambda$. Hence, by substituting the value of x and y into the constraint equation, we obtain that $\lambda = -\frac{1}{2}$. And this gives us the solution as $x = y = \frac{1}{2}$. Hence, at point $P(0.5, 0.5)$ the function is minimized. Also note that point P is not the absolute minimum of the function $f(x, y)$. The absolute minimum is at (0, 0). But the conditional minimum is at P.

1.2 Vectors

In this section, we shall briefly present the key aspects of vector algebra, vector calculus that are later needed to derive some of the results of the book. It should also be noted that vector analysis is the standard mathematical tool and the mode of thought in modern physics and especially in classical electromagnetism.

1.2.1 Vector algebra

Scalars and *vectors* represent the vast majority of the physical quantities. Scalars are described by real numbers and the description of a scalar in this way gives the complete specification about the physical quantity. However, there are certain quantities called 'vectors' that require the specification of the direction for them to make any physical sense and to complete their description. Scalars obey the rules of ordinary algebra and ordinary calculus. Vectors, on the other hand, must obey the *parallelogram law of vector additions* (described later). Examples of scalars include mass, temperature, energy, etc. Examples of vectors include velocity, acceleration, force, etc. Scalars are represented by real numbers, while vectors are shown by directed line elements in space, as shown in figure 1.3.

By convention, vectors are indicated by a small arrow over the letter that represents the vector (e.g.) or by bold face characters (e.g. **A**). Furthermore, the points representing the head and the tail end of a vector $\vec{A}$ are taken to be O and A, respectively. Scalars, on the other hand, are represented by the alphabet (e.g. temperature—T).

The equality of two vectors

We say that two vectors are equal if they have the same magnitude and direction and represent the same physical quantity. With such a definition, we can see that a parallel translation of a vector without changing its magnitude or direction does not bring any change to the vector. If a vector is multiplied by a non-zero positive constant c, its direction remains the same, but the magnitude is multiplied by c. If, on the other hand, a negative non-zero quantity multiplies a vector, the magnitude of that quantity multiplies the magnitude of the vector, while its direction flips by 180°. If we multiply a vector by zero, in order to have mathematical consistency, we

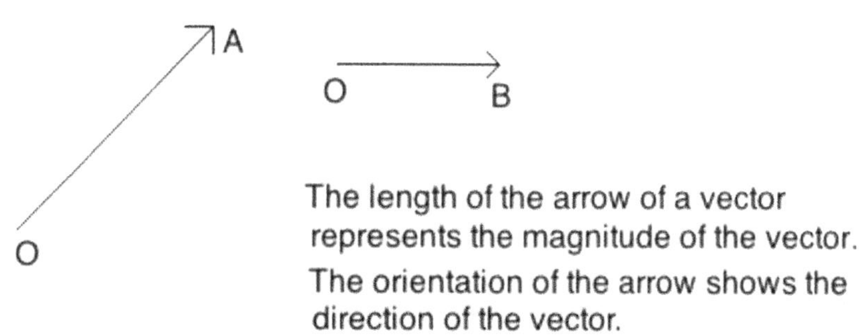

Figure 1.3. Directed line segments are used to show vectors in physics.

obtain something called the 'zero vector'. It is a vector of zero magnitude and its direction is indeterminate.

Parallelogram law of vector addition

Two vectors $\vec{A}$ and $\vec{B}$ with the same dimensions can be added to form a third vector $\vec{C}$. If the tail end O of $\vec{A}$ coincides with the head B of $\vec{B}$, the vector $\vec{C} = \vec{A} + \vec{B}$ is obtained by joining the tail of $\vec{A}$ with the head of $\vec{B}$, as shown in figure 1.4. This is called the 'triangle law of vector addition'.

Suppose that the magnitude of $\vec{A}$ is A and that the magnitude of $\vec{B}$ is B, what is the magnitude and the direction of $\vec{C}$? In order to calculate this, we rearrange the vectors such that the tail ends of the vectors meet, as shown in figure 1.5. Let θ be the angle between the vectors. The vector **A** is along the line PQ and vector **B** is along the line QS. The sum of the vectors **C** is along the line PS. We construct a parallelogram, as shown in figure 1.5.

Let the T be the perpendicular point from S on the line PQ. And, if θ is the angle between the vectors, we obtain from the diagram that

$$PS^2 = (PQ + QT)^2 + ST^2$$
$$\Rightarrow \mathbf{C}^2 = (\mathbf{A} + \mathbf{B}\cos\theta)^2 + (\mathbf{B}\sin\theta)^2.$$

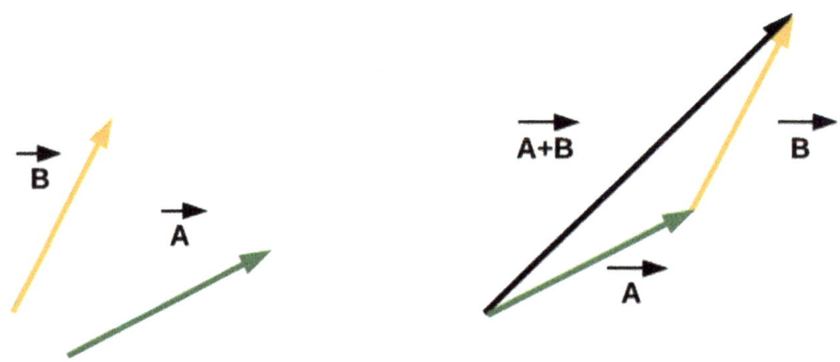

Figure 1.4. A diagram to demonstrate the triangle law of addition of vectors $\vec{A}$ and $\vec{B}$.

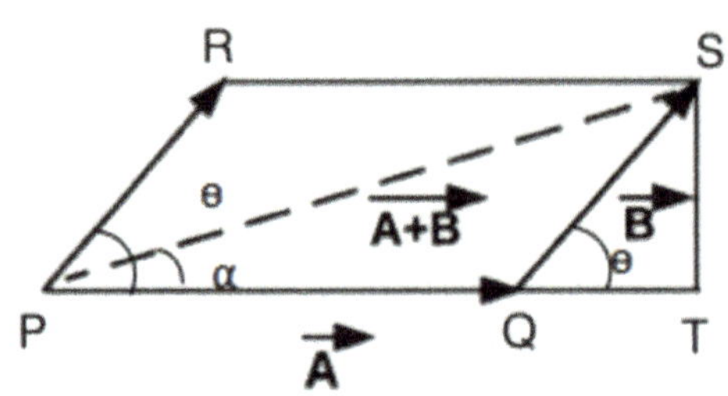

Figure 1.5. A diagram to illustrate the parallelogram law of the addition of vectors. A vector **A** is represented by the line segment PQ and another vector **B** is represented by the line segment QS. PS is the resultant vector **C** obtained by the addition of **A** and **B**.

Thus the magnitude of (**A** + **B**) is given by the following formula

$$\left|\vec{A}+\vec{B}\right| = \sqrt{\left|\vec{A}\right|^2 + \left|\vec{B}\right|^2 + 2\left|\vec{A}\right|\left|\vec{B}\right| \cos\theta}.$$

And, if its angle with **A** is α, this is given by

$$\tan\alpha = \frac{ST}{PT} = \frac{\left|\vec{B}\right| \sin\theta}{\left|\vec{A}\right| + \left|\vec{B}\right| \cos\theta}$$

The parallelogram law of vector addition obeys the commutative rule:

$$\vec{A} + \vec{B} = \vec{B} + \vec{A}.$$

The proof of this can easily be shown with the help of figure 1.6.

Subtraction of vectors

Let **A** and **B** be two vectors. We define (**A** – **B**) as the sum of vectors **A** and (–**B**). Hence to subtract **B** from **A**, invert the direction of **B** and add it to **A**, as shown in figure 1.7.

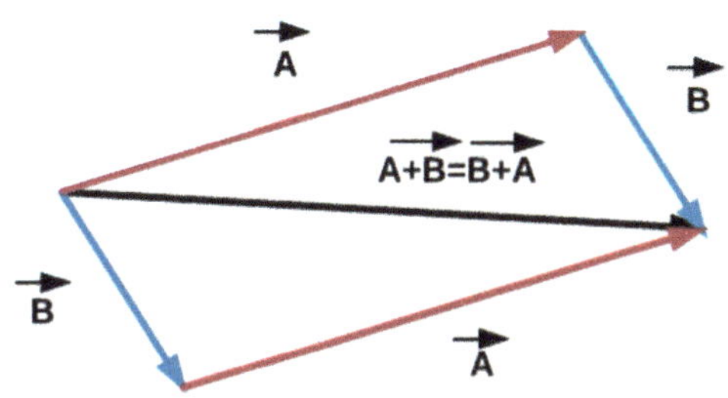

Figure 1.6. The two equivalent ways of adding the two vectors.

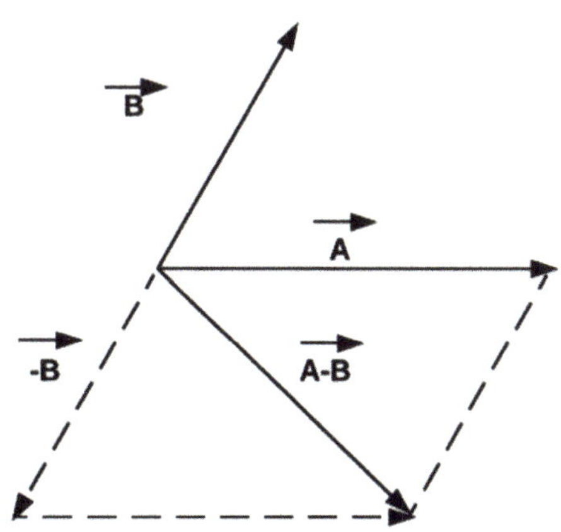

Figure 1.7. A geometrical diagram to demonstrate the subtraction of two vectors.

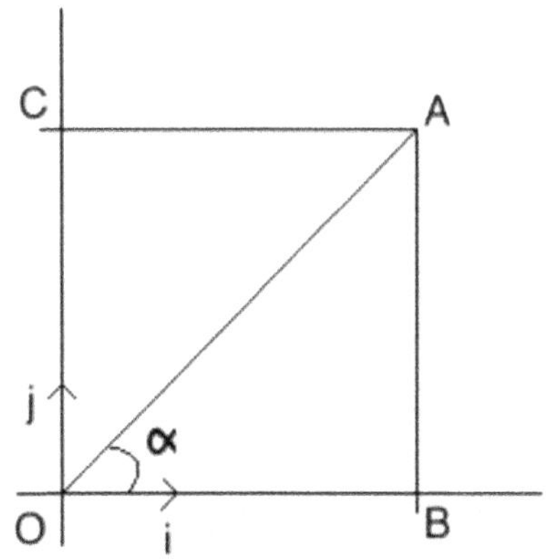

Figure 1.8. Perpendiculars AB and AC are drawn from the vector **OA** to resolve it into its corresponding x and y components.

Resolution of vectors

Consider a vector **A** drawn from the origin O to point A, as shown in figure 1.8.

Let the angle the vector **A** makes with the x-axis be α. Let us now draw the perpendiculars AB and AC from point A on the x- and y-axes, respectively. Using the identities of trigonometry, we have from the triangle law of vector addition that

$$\vec{A} = \left|\vec{A}\right| \cos\alpha\hat{\imath} + \left|\vec{A}\right| \sin\alpha\hat{\jmath}.$$

Here $|\vec{A}|$ represents the magnitude of the vector **A**, and $\hat{\imath}$ and $\hat{\jmath}$ denote the vectors of unit magnitude along the x- and y-axes, respectively. We usually write the vector **A** as follows:

$$\vec{A} = A_x\hat{\imath} + A_y\hat{\jmath},$$

where $A_x = |\vec{A}|\cos\alpha$ and $A_y = |\vec{A}|\sin\alpha$.

If **A** is a three-dimensional vector, we will have an additional component along the z-axis and a unit vector along the z-axis represented by $\hat{k}$.

Scalar product of two vectors

There are two ways in which two vectors can be multiplied. The first method of multiplying two vectors is called the dot or scalar product. Suppose that we have two vectors **A** and **B** with magnitudes $|\vec{A}|$ and $|\vec{B}|$ and an angle ϕ between them, as shown in figure 1.9.

The scalar product is defined as

$$\vec{A} \cdot \vec{B} = \left|\vec{A}\right|\left|\vec{B}\right| \cos\phi.$$

Since we know that the x- and y-axes are perpendicular to each other, we have that

$$\hat{\imath} \cdot \hat{\jmath} = 1 \times 1\cos 90^\circ = 0.$$

Also, since the angle between the x-axis and x-axis or y-axis and y-axis is 0°, we have that

$$\hat{\imath} \cdot \hat{\imath} = \hat{\jmath} \cdot \hat{\jmath} = 1 \times 1\cos 0^\circ = 1.$$

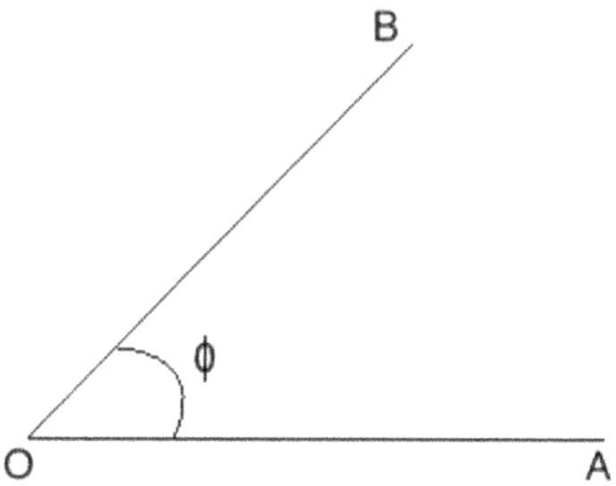

Figure 1.9. Two vectors **A** (represented as **OA**) and **B** (represented as **OB**) with an angle ϕ between them.

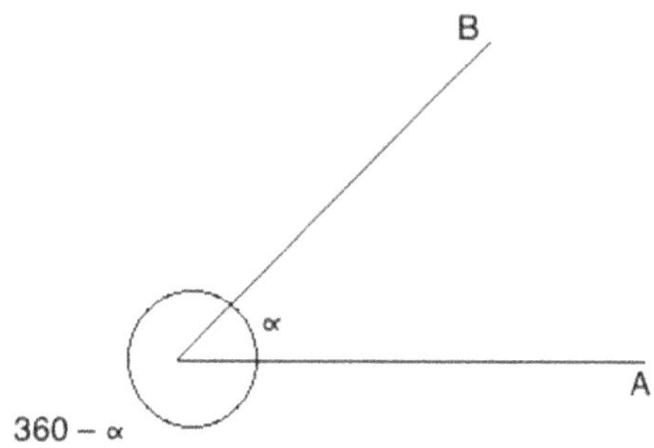

Figure 1.10. In the vector cross product between two vectors **A** and **B** we only have to consider the smaller angle α, as shown in this figure.

With these definitions, let us define the scalar product in terms of components of the two vectors. We have that

$$\vec{A} \cdot \vec{B} = \left(A_x\hat{i} + A_y\hat{j}\right) \cdot \left(B_x\hat{i} + B_y\hat{j}\right) = A_xB_x + A_yB_y.$$

Note that the scalar product of two vectors is unique and gives a scalar. We shall see another product of two vectors that gives a unique vector.

Vector product of two vectors

The vector product of two vectors **A** and **B** is a vector and is denoted as $\vec{A} \times \vec{B}$. The magnitude of the cross product $\vec{A} \times \vec{B}$ is defined as follows:

$$\vec{A} \times \vec{B} = \left|\vec{A}\right|\left|\vec{B}\right| \sin\alpha,$$

where $|\vec{A}|$ and $|\vec{B}|$ represent the magnitude of vectors $\vec{A}$ and $\vec{B}$, respectively, and α is the smaller angle between them, as shown in figure 1.10.

The direction of $\vec{A} \times \vec{B}$ is perpendicular to both $\vec{A}$ and $\vec{B}$. To determine the direction of the cross product, we use the right-hand thumb rule: draw the vectors $\vec{A}$ and $\vec{B}$ with tail ends coinciding at a point O. Next place your stretched right palm perpendicular to the plane in such a way that the fingers are along the vector $\vec{A}$ and when the fingers are closed they go toward vector $\vec{B}$. Then the direction of $\vec{A} \times \vec{B}$ is given by the direction of the thumb. This is shown in figure 1.11.

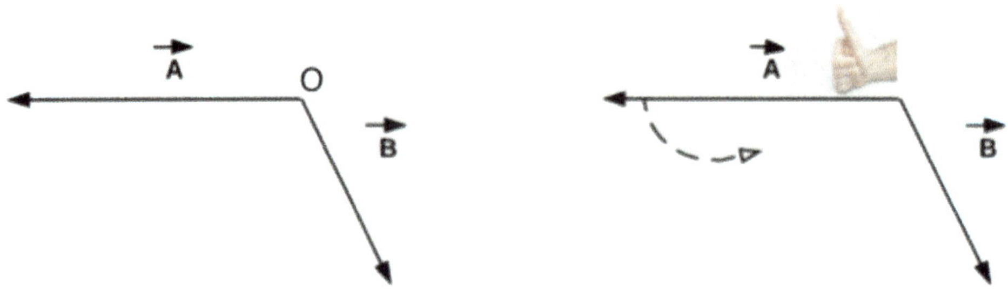

Figure 1.11. A way to determine the direction of the cross product $\vec{A} \times \vec{B}$ of two vectors lying in a plane.

Then from the definition of the cross product, we have that

$$\vec{A} \times \vec{B} = -\vec{B} \times \vec{A}.$$

Let us now try to understand what a right-handed Cartesian coordinate system means. Once we have specified that the x- and y-axes are perpendicular to each other, there are two ways of choosing the z-axis. We can either choose the z-axis so that

$$\hat{i} \times \hat{j} = \hat{k} \qquad \text{(right-handed system)}$$

or choose it so that

$$\hat{i} \times \hat{j} = -\hat{k} \qquad \text{(left-handed system)}$$

Unless otherwise specified, we shall be assuming that the coordinates in this book are right-handed. Hence, we have that

$$\begin{aligned} \hat{i} \times \hat{j} &= \hat{k} \\ \hat{j} \times \hat{k} &= \hat{i} \\ \hat{k} \times \hat{i} &= \hat{j}. \end{aligned}$$

Also, by the definition of the cross product, we have that

$$\hat{i} \times \hat{i} = 1 \times 1 \sin 0^\circ = 0.$$

Similarly, we have that $\hat{j} \times \hat{j} = \hat{k} \times \hat{k} = 0$. With these definitions, we obtain that

$$\vec{A} \times \vec{B} = \left(A_x\hat{i} + A_y\hat{j} + A_z\hat{k}\right) \times \left(B_x\hat{i} + B_y\hat{j} + B_z\hat{k}\right)$$
$$\Rightarrow \vec{A} \times \vec{B} = \left(A_yB_z - B_yA_z\right)\hat{i} + \left(A_zB_x - B_zA_x\right)\hat{j} + \left(A_xB_y - B_xA_y\right)\hat{k}.$$

1.2.2 Vector calculus

Suppose that a vector **A** varies with time t so that **A** = **A**(t). The rate of change of **A** with respect to time, similarly to scalars, is defined as

$$\frac{\mathrm{d}\vec{A}}{\mathrm{d}t} = \lim_{\delta t \to 0}\left[\frac{\vec{A}(t + \delta t) - \vec{A}(t)}{\delta t}\right].$$

Note, however, that $\frac{d\vec{A}}{dt}$ is another vector and can be represented in terms of components just like any other vector.

We also have that for two vectors **A** and **B**,

$$\frac{d}{dt}\left(\vec{A}\cdot\vec{B}\right)=\frac{d\vec{A}}{dt}\cdot\vec{B}+\vec{A}\cdot\frac{d\vec{B}}{dt}$$

$$\frac{d}{dt}\left(\vec{A}\times\vec{B}\right)=\frac{d\vec{A}}{dt}\times\vec{B}+\vec{A}\times\frac{d\vec{B}}{dt}.$$

It is evident that the laws of vector calculus are analogous to those in ordinary calculus, but with the difference that the order of the vectors in the cross product matters.

Vector line integrals

Consider a function $f(x, y)$ of two independent variables x, y. Suppose that the points P and Q are defined as the end points of the domain of f. Now, we want to find out the integral S along the path defined by l, as shown in figure 1.12(*a*).

The line integral is written as

$$S=\int_{P}^{Q} f(x, y)\mathrm{d}l$$

In order to evaluate the integral S, we move along the path from P to Q and find out the value of the function f at each point. The graph of f against the length of the curve is plotted. Let us suppose that it is as shown in figure 1.12(*b*).

Then, we have that S = the area under the f–l curve from P to Q. Let us illustrate this with a simple example.

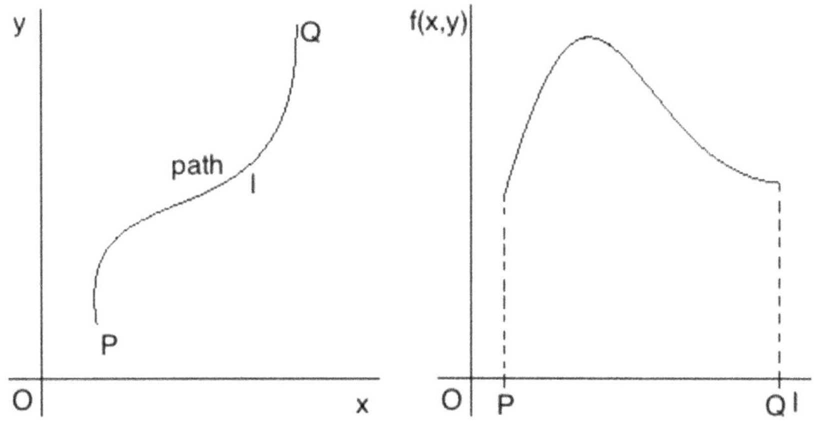

Figure 1.12. (*a*) The graph of f as a function of the two independent variables x and y. (*b*) The graph of f against the path l.

Example Consider the function $f(x, y) = xy$ defined for all integers. And suppose we need to evaluate the following integral S along the path given by $x = y$. This integral is written as

$$S = \int_{(0,0)}^{(1,1)} f(x, y)\mathrm{d}l.$$

Solution

Now, consider an infinitesimal length element $\mathrm{d}l$ as shown in figure 1.13. By Pythagoras' theorem, we need to have that

$$(\mathrm{d}l)^2 = (\mathrm{d}x)^2 + (\mathrm{d}y)^2. \quad (1.2)$$

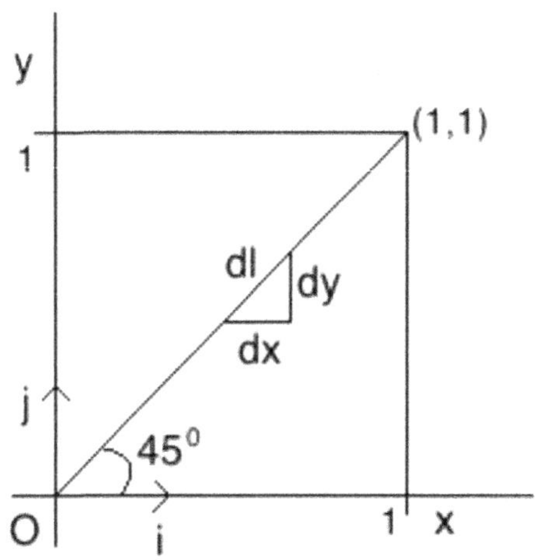

Figure 1.13. The infinitesimal element of the path l is shown and its connection with the infinitesimal $\mathrm{d}x$ and $\mathrm{d}y$ is apparent from Pythagoras' theorem.

Since, we are on the line $x = y$, we know that $\mathrm{d}x = \mathrm{d}y$. Hence, equation (1.2) becomes

$$\mathrm{d}l = \sqrt{2}\,\mathrm{d}x.$$

It is also evident that as we move from (0, 0) to (1, 1) along the line l, x varies from 0 to 1. Hence, we have

$$S = \sqrt{2}\int_0^1 x^2\,\mathrm{d}x = \frac{\sqrt{2}}{3}.$$

Next, suppose that we have a vector **A** and need to evaluate its line integral along the vector **dl**, we write

$$S = \int_P^Q \vec{A} \cdot \mathrm{d}\vec{l} = \int_P^Q \left(A_x \mathrm{d}l_x + A_y \mathrm{d}l_y + A_z \mathrm{d}l_z\right).$$

The individual integrals are evaluated just like the normal scalar line integrals, as we saw earlier.

Gradient of a scalar field

For a one-dimensional function $f(x)$ of only one independent variable x, the gradient is defined as the slope of the function at a particular point. Let us now extend this concept to a three-dimensional scalar function $g(x, y, z)$. The gradient of this function is denoted as ∇g and is defined in Cartesian coordinates as

$$\nabla g = \frac{\partial g}{\partial x}\hat{i} + \frac{\partial g}{\partial y}\hat{j} + \frac{\partial g}{\partial z}\hat{k}.$$

Hence, the gradient of a scalar function is a vector. The gradient of a scalar function points in the direction of the greatest rate of increase of that function and its magnitude is the slope of the graph in that direction. If dg represents the change in the function along the path dl, we have by the definition of the gradient that

$$\mathrm{d}g = \nabla g \cdot \mathrm{d}\vec{l}.$$

Next, suppose that a function T has the property that d$T = 0$ for some **dl**. We then have that

$$\mathrm{d}T = \nabla T \cdot \mathrm{d}\vec{l} = 0. \tag{1.3}$$

This implies that ∇T and $\mathrm{d}\vec{l}$ are perpendicular to each other. If T represents temperature, we see that the paths given by **dl** represent isotherms. Next, we have that

$$\int_P^Q \mathrm{d}T = T(Q) - T(P). \tag{1.4}$$

We can make use of equation (1.3) to rewrite equation (1.4) as

$$\int_P^Q \mathrm{d}T = \int_P^Q \nabla T \cdot \mathrm{d}\vec{l} = T(Q) - T(P).$$

It is clear that the line integral is independent of the path taken. Hence $\int_P^Q \nabla T \cdot \mathrm{d}\vec{l}$ is independent of the path taken.

Suppose we have a line integral of a general vector **A** independently of the path taken, we say that **A** is a *conservative field*. Since the line integral of the gradient of a scalar function is independent of the path taken, we can say that a general conservative field can be written as the gradient of a scalar function. Hence, if we can write a general vector **A** as a gradient of a scalar function φ, we can say that **A** is a conservative field.

Suppose that $\vec{A} = \nabla\phi$. Let us try to evaluate the following integral:

$$\oint \vec{A} \cdot \mathrm{d}\vec{l}.$$

Since **A** is a conservative field and only depends on the end points of the integration, the above integral should be zero. Hence, we have that

$$\oint \vec{A} \cdot \mathrm{d}\vec{l} = \oint (\nabla\phi) \cdot \mathrm{d}\vec{l} = 0. \tag{1.5}$$

A typical physical example of a conservative field is the gravitational force field. The line integral of such a force field represents the work done in moving the particle in the field and it is zero if the particle's displacement is zero, as shown by equation (1.5).

Divergence of a vector field

The divergence of a vector field **A** is defined as

$$\nabla \cdot \vec{A} = \lim_{V_i \to 0} \frac{\oint_{S_i} \vec{A} \cdot \mathrm{d}\vec{S}_i}{V_i}.$$

Here, S_i represents the closed surface area and V_i represents the volume enclosed by S_i. In Cartesian coordinates, we have

$$\nabla \cdot \vec{A} = \frac{\partial A_x}{\partial x} + \frac{\partial A_y}{\partial y} + \frac{\partial A_z}{\partial z}.$$

It is fairly evident that the divergence is a scalar quantity. Now, if we assume the 'nabla' operator (∇) to be a vector, we can easily remember that the gradient of a scalar function f is defined as

$$\nabla f = \left(\frac{\partial}{\partial x}\hat{i} + \frac{\partial}{\partial y}\hat{j} + \frac{\partial}{\partial z}\hat{k}\right) f.$$

And with the definition of the scalar product, the divergence of a vector function can be conveniently written as

$$\nabla \cdot \vec{A} = \left(\frac{\partial}{\partial x}\hat{i} + \frac{\partial}{\partial y}\hat{j} + \frac{\partial}{\partial z}\hat{k}\right) \cdot \vec{A}.$$

Every vector field can be visualized as being represented by field lines. If a vector has a net positive divergence at a particular point P in space, this means that at P the field lines are directed away from the point. If it has negative divergence at a point Q, it means that the field lines are directed toward the point Q. This is the physical significance of the divergence.

Curl of a vector field
Consider a vector **A** with the curl of **A** being defined in Cartesian coordinates as

$$\nabla \times \vec{A} = \left(\frac{\partial}{\partial x}\hat{i} + \frac{\partial}{\partial y}\hat{j} + \frac{\partial}{\partial z}\hat{k} \right) \times \vec{A}.$$

If a vector field has a non-zero curl, this means that it has circulation or vorticity. Suppose that we have a vector field **A** and it has a non-zero curl, meaning that $\nabla \times \vec{A} \neq \vec{0}$, then we know that the velocities in this field have a component of circular rotation on a general flow in one direction. For instance, the velocities of water particles flowing out of a bathtub from a drain generally acquire a curl in one particular direction. A vector field **B** with $\nabla \times \vec{B} = \vec{0}$ everywhere is said to be an *irrotational* vector field.

The Laplacian operator
So far we have seen that the gradient of a scalar field φ gives us a vector field and that the divergence of a vector function **A** gives us a scalar field. Now, there are two ways we can form new unique fields. We can either take the divergence of the gradient of a scalar field φ and form a new scalar field or take the gradient of the divergence of the vector field **A** and form a new vector field. The former operation is very important in electrodynamics, and in physics generally, and we shall study this now.

Consider a scalar field φ defined everywhere in space. The gradient of this field is given in Cartesian coordinates as

$$\nabla \varphi = \frac{\partial \varphi}{\partial x}\hat{i} + \frac{\partial \varphi}{\partial y}\hat{j} + \frac{\partial \varphi}{\partial z}\hat{k}.$$

Next, take the divergence of the above vector field to give

$$\nabla \cdot (\nabla \varphi) = \frac{\partial^2 \varphi}{\partial x^2} + \frac{\partial^2 \varphi}{\partial y^2} + \frac{\partial^2 \varphi}{\partial z^2}.$$

The above equation can be written as

$$\nabla^2 \varphi = \left(\frac{\partial^2}{\partial x^2} + \frac{\partial^2}{\partial y^2} + \frac{\partial^2}{\partial z^2} \right) \varphi.$$

Hence, we define the new scalar differential operator as

$$\nabla^2 = \left(\frac{\partial^2}{\partial x^2} + \frac{\partial^2}{\partial y^2} + \frac{\partial^2}{\partial z^2} \right).$$

This operator is called the *Laplacian* operator. It acts on scalar functions and gives us another scalar function. In order to see the significance of the operator, let us reduce the field to a one-dimensional field $\varphi(x)$ dependent on x. The Laplacian in this case would simply be

$$\nabla^2\varphi(x) = \frac{\partial^2\varphi(x)}{\partial x^2}.$$

We have seen before that if $\nabla\varphi(x) = 0$ and $\nabla^2\varphi(x) > 0$ at a point x_1, this implies that x_1 is a point of local minimum of $\varphi(x)$ and if $\nabla\varphi(x) = 0$ and $\nabla^2\varphi(x) < 0$ at a point x_2, this implies that x_2 is a point of local maximum of $\varphi(x)$.

The divergence or Gauss theorem

Suppose that we have a vector field **A** defined everywhere and let us consider a volume V surrounded by a closed surface S. Then the Gauss or divergence theorem states that

$$\oint_S \vec{A} \cdot \mathrm{d}\vec{S} = \int_V \left(\nabla \cdot \vec{A}\right)\mathrm{d}V.$$

Suppose that the divergence of the vector **A** is zero. We then have that for any closed surface S

$$\oint_S \vec{A} \cdot \mathrm{d}\vec{S} = 0.$$

Let us now divide the closed surface S into two open surfaces S_1 and S_2 joined at the rim R. We have taken the directions of the surfaces S_1 and S_2 as shown in figure 1.14.

Hence, we can write the above equation as

$$\oint_S \vec{A} \cdot \mathrm{d}\vec{S} = \int_{S_1} \vec{A} \cdot \mathrm{d}\vec{S} - \int_{S_2} \vec{A} \cdot \mathrm{d}\vec{S} = 0. \tag{1.6}$$

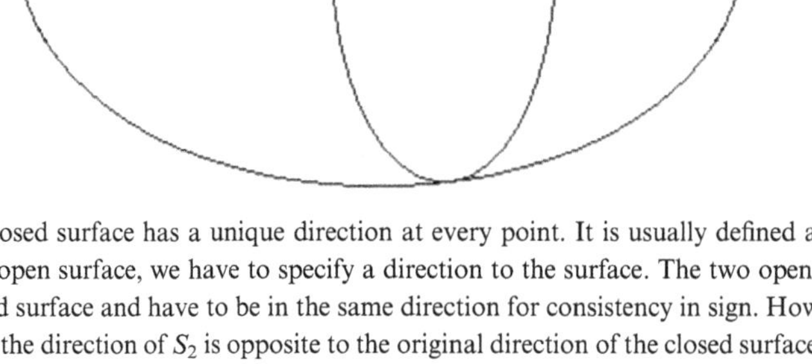

Figure 1.14. A closed surface has a unique direction at every point. It is usually defined as outward normal. However, for an open surface, we have to specify a direction to the surface. The two open surfaces S_1 and S_2 constitute a closed surface and have to be in the same direction for consistency in sign. However, we can see in this diagram that the direction of S_2 is opposite to the original direction of the closed surface S. Hence, we need to put a minus sign in equation (1.6).

There is a minus sign in front of the second integral because we have reversed the direction of the open surface S_2 from its previous direction, as shown in figure 1.14. Hence, we have that

$$\int_{S_1} \vec{A} \cdot d\vec{S} = \int_{S_2} \vec{A} \cdot d\vec{S}. \tag{1.7}$$

So, as seen from equation (1.7), if the divergence of a vector field is zero, the surface integral depends only on the rim of the surface and not on the actual surface chosen. On the other hand, if the divergence of a vector field is not zero, the surface integral depends both on the rim and the surface. When the divergence of a vector field is zero, it is called a *solenoidal* vector field.

Stokes' theorem

Another important theorem in vector calculus is Stokes' theorem. Suppose we have a vector field **A** and a non-planar surface S bounded by a rim C. Then, Stokes' theorem states that

$$\oint_C \vec{A} \cdot d\vec{l} = \int_S \left(\nabla \times \vec{A}\right) \cdot d\vec{S}.$$

One important result immediately follows from Stokes' theorem: consider two open surfaces S_1 and S_2, which share the same rim C. Stokes' theorem implies that we have

$$\int_{S_1} \left(\nabla \times \vec{A}\right) \cdot d\vec{S} = \int_{S_2} \left(\nabla \times \vec{A}\right) \cdot d\vec{S}.$$

We can write the above equation as

$$\int_{S_1} \left(\nabla \times \vec{A}\right) \cdot d\vec{S} - \int_{S_2} \left(\nabla \times \vec{A}\right) \cdot d\vec{S} = 0. \tag{1.8}$$

Hence, we can form a closed surface S with volume V that is bounded by S_1 and S_2 and it follows from equation (1.8) that

$$\oint_S \left(\nabla \times \vec{A}\right) \cdot d\vec{S} = 0.$$

However, we have from the divergence or Gauss theorem that

$$\oint_S \left(\nabla \times \vec{A}\right) \cdot d\vec{S} = \int_V \left[\nabla \cdot \left(\nabla \times \vec{A}\right)\right] dV = 0.$$

Since this is valid for any closed surface S containing any volume, we have for any general vector **A**

$$\nabla \cdot \left(\nabla \times \vec{A}\right) = 0.$$

So, we arrive at an important result that $\nabla \times \vec{A}$ is a solenoidal vector field. Next, we have seen before that for a conservative field **B**, the line integral along any closed path C is zero. Hence, we have that

$$\oint_C \vec{B} \cdot \mathrm{d}\vec{l} = 0.$$

But, from Stokes' theorem we have for any open surface S attached to the path C,

$$\oint_C \vec{B} \cdot \mathrm{d}\vec{l} = \int_S \left(\nabla \times \vec{B}\right) \cdot \mathrm{d}\vec{S} = 0.$$

Since this is generally true of any surface, we have that

$$\left(\nabla \times \vec{B}\right) = \vec{0}.$$

And if **B** is a conservative field, it can be written as the gradient of a scalar potential φ and hence we have a general result that

$$\nabla \times (\nabla\varphi) = \vec{0}.$$

Thus, a conservative field is an irrotational field. Finally, it can be proven that, in Cartesian coordinates, for a general vector field **A**

$$\nabla \times \left(\nabla \times \vec{A}\right) = \nabla\left(\nabla \cdot \vec{A}\right) - \nabla^2\vec{A}.$$

Conclusion

We have learnt how to find out the absolute and conditional extremum of functions and have familiarized ourselves with various vector analysis tools that will be used later in the book to represent various electromagnetic phenomena.

References

[1] Jaynes E T 1980 The minimum entropy production principle *Annu. Rev. Phys. Chem.* **31** 579
[2] Maxwell J C 1954 *A Treatise on Electricity and Magnetism* (New York: Dover) pp 407–8
[3] Feynman R P, Leighton R B and Sands M 1964 *The Feynman Lectures on Physics* vol 2 (Reading, MA: Addison-Wesley) ch 19

IOP Publishing

N R Sree Harsha, Anupama Prakash and D P Kothari

Chapter 2

The Concept of charge

2.1 Electric charge

Classical electromagnetism deals with the properties of electric charges and currents and their interaction with unlimited precision. There are two types of electric charges. One type of charge, by convention, is called 'positive charge' and the other type is called 'negative charge'. It is known through experiments that like charges always repel each other and unlike charges always attract each other. The force of the attraction and repulsion depends on the quantity of the charge and reduces as the square of the distance between the particles. It is often remarked that gravitational forces act similarly to electromagnetic forces, but with two striking differences: there is only one type of gravitational 'charge', represented by its 'mass', and electromagnetic forces are a billion-billion-billion-billion times stronger than gravitational forces. Now, a question that might occur to an inquisitive mind is this: how can you compare gravitational forces and electromagnetic forces? Are they not the result of totally different properties of particles? In order to understand this, we must first know a little bit about the standard model of particle physics.

The study of fundamental particles like protons, neutrons, electrons, neutrinos, etc is called high-energy physics or elementary particle physics. The Greek philosopher Democritus postulated that matter is ultimately made up of indivisible particles and he called these 'atoms' ('atom' means 'indivisible'). Of course, we now know through Rutherford's experiments that an atom consists of a nucleus at its center with electrons revolving around it and we break up atoms all the time in laboratories. The central nucleus consists of protons and neutrons (together called 'nucleons').

With the advances made in particle accelerators, we further discovered that protons and neutrons are made of even more fundamental particles called 'quarks'. Particle physicists now believe that they can describe all known matter within a single framework called the *standard model of particle physics*. This states that there exists a very simple scheme of two sets of particles called 'quarks' and 'leptons' that make up all the matter around us. The forces between these particles, such as

doi:10.1088/978-0-7503-1266-0ch2

electromagnetic, strong and weak nuclear forces, are themselves described by the exchange of particles called 'gauge bosons'. An example of such a gauge boson that transmits electromagnetic forces is the 'photon' (the quantum of light). Note that the standard model does not yet describe the force of gravity.

These fundamental particles form various combinations, which we observe to be protons, neutrons, etc. So, we know that matter is composed of tiny particles called quarks and leptons.

- Quarks come in six different varieties: up (u), down (d), charm (c), strange (s), top (t) and bottom (b). Quarks also have their respective anti-matter counterparts, which are represented by a line over the quark letter symbol. Now, different flavors of quarks combine to form 'baryons' and quarks and anti-quarks combine to form 'mesons'. Protons and neutrons are examples of baryons. Negative and positive kaons are examples of mesons.
- Electrons, muons, taus and neutrinos form another class of particles called *leptons*.

Quarks and leptons have an intrinsic angular momentum called spin, which is equal to a half-integer (1/2) of the basic unit, and are called *fermions*. Particles that have zero or integer spin are called *bosons* (for example, gauge bosons).

Returning to our previous question, we can now meaningfully compare the gravitational force and the electric force felt by a fundamental particle such as the electron. We see that the ratio of the magnitude of the forces is

$$\left.\frac{F_{\text{gravity}}}{F_{\text{elec}}}\right|_{\text{electron}} \approx 2.4 \times 10^{-43}$$

It is, however, instructive to know what Feynman [1] said on this topic.

> *People who want to make electricity and gravity out of the same thing will find that electricity is so much more powerful than gravity, it is hard to believe they could both have the same origin. How can I say one thing is more powerful than another? It depends on how much charge you have and how much mass you have. You cannot talk about how strong gravity is by saying: 'I take a lump of such a size', because you chose the size. If we try to get something that Nature produces—her own pure number that has nothing to do with inches or years or nothing to do with our own dimensions—we can do it this way. If we take a fundamental particle such as an electron—any different one will give a different number, but to give an idea say electrons—two electrons are two fundamental particles, and they repel each other inversely as the square of distance due to electricity, and they attract each other as the square of the distance due to gravitation. What is the ratio of the gravitational force to the electrical force? The ratio of gravitational attraction to electrical repulsion is given by a number with 42 digits tailing off.*

We have seen that charge is the fundamental property of particles in the standard model. The charge of a particle can never be isolated from its mass. We also know that like charges repel each other and unlike charges attract each other. If we have a pool of positively and negatively charged particles, after sufficient time we must expect that the number of positive and negative charges will be equal for a given volume V. They try to mix to form atoms in such a way that the net electrical forces are cancelled out. The protons and neutrons form nuclei and electrons revolve around the nuclei. But the uncertainty principle of quantum mechanics dictates how close an electron can get to the nucleus of the atom. If we try to confine an electron in a region very close to a proton, the electron will have a very large mean square momentum and this grows larger the more we try to confine it. Hence the electrons in an atom cannot get very close to the nucleus. So, for equilibrium we must expect the positive charge to be created at the center of an atom and electrons revolving around it. This makes an atom electrically neutral overall. This simple picture has a deep meaning hidden in it. By default, it assumes that the charge of the electrons (or fundamental particles in general) does not depend on their velocity. *This fact is not at all obvious* a priori.

An experiment to identify the dependence of the charge of the electron on its velocity was performed by D F Bartlett and B F L Ward [2]. They assumed that an electron's charge varies slightly with its velocity. They also assumed that the charge of an electron is e when it is at rest with respect to us. And its charge is q when it is moving with a finite velocity v with respect to us. If c represents the speed of light in vacuum and k represents a dimensionless constant, they proposed that

$$q = e\left[1 + \frac{kv^2}{c^2}\right] \quad \text{where } |k| \ll 1.$$

They tested this hypothesis for conduction electrons in a current-carrying solenoid inside a Faraday ice pail and concluded that $|k| < 2 \times 10^{-15}$. We shall, however, assume that $k = 0$. This means that the quantity 'charge' is a scalar and invariant under linear coordinate transformation. Thus, in our model, the atom is electrically neutral overall.

In the nucleus there are several protons and neutrons and they are held together in a very small volume. Since protons have a positive charge, they should repel each other. What holds the nucleus together? In nuclei there are also non-nuclear forces called the nuclear forces that bind the nucleus together. These nuclear forces are short-range forces and decrease much more rapidly than $1/r^2$.

Consider a fluorine atom, which has an equal number of electrons and protons—nine each—making the atom neutral. If an electron is added, the number of electrons in the atom exceeds the number of protons, and the atom then acquires a negative charge.

Thus, the excess of electrons results in a negative charge.

We have seen that the addition of electrons to a neutral atom negatively charges it. Similarly, can we charge an atom positively by adding a proton? Not at all. Protons cannot be transferred and cannot exist independently in the way electrons can. This is because they are tightly packed in the nucleus. It is practically impossible to isolate protons from the nucleus. How, then, does a body acquire a positive charge?

Suppose we remove an electron from a neutral fluorine atom. Then the number of electrons is eight, which is less than the number of protons (nine), leaving the atom positively charged.

Thus, the deficiency of the electrons results in a positive charge.

2.2 Electrification

The phenomenon of charging a neutral body is called *electrification.* As we have seen, the outermost electrons of some atoms can be easily removed from the bounds of the nucleus and they tend to have high mobility. And there are three ways of charging material objects, as follows.

Frictional electrification

Different materials are made of different types and combinations of atoms. Hence, they have different chemical and electrical properties. One such property is known as 'electron affinity'. The atoms of a material with a high electron affinity can attract electrons more easily than a material with a lower electron affinity. Consider the example of rubbing a glass rod and silk. When they are rubbed against each other, they are forced into close proximity. The atoms with a higher electron affinity attract the free electrons of the other material, resulting in the creation of positive and negatively charged bodies. Thus, silk, which has greater electron affinity, attracts electrons from the glass rod and becomes negatively charged. The glass rod is then deficient of electrons and becomes positively charged. This process is called 'tribo-electric charging'. The triboelectric effect (also known as triboelectric charging) is a type of contact electrification in which certain materials become electrically charged after they come into contact with a different material [3]. Rubbing glass with fur, or running a comb through the hair, can build up triboelectricity. Most everyday static electricity is triboelectric and this is a particular concern for the electronics industry. Static charge build-up can lead to detrimental effects for semiconductor devices upon discharge and such discharge is more specifically called as electrostatic discharge (ESD).

This process of charging bodies by friction is called frictional electrification.

John Carl Wilcke published the first triboelectric series in a 1757 paper on static charges. The different types of materials are listed in the series according to their electron affinities, as shown in figure 2.1. In the series, the materials towards the bottom of the series, when rubbed against a material near the top of the series, will acquire a more negative charge since they have greater electron affinity. It can also be understood that the greater the charge transferred between two given materials, the further apart they are in the series. Hence, to find out whether a material object will become positively or negatively charged when rubbed against another object, we can check its relative position in the triboelectric series.

Figure 2.1. The triboelectric series for a few objects seen everyday.

Conduction

The process of charging a body by placing it in contact with another charged body is called charging by *conduction*. Let us demonstrate this method by means of a simple example of charging a neutral sphere using a negatively charged ebonite rod. Consider an uncharged sphere *A* on an insulated stand, as shown in figure 2.2. Let us now bring a negatively charged ebonite rod into contact with the neutral sphere. As a result of this action, the neutral sphere obtains a negative charge.

The key to understanding charging by conduction is to understand that like charges repel and try to spread out as far away as possible. The negatively charged rod has an excess of electrons and they repel each other and hence are present on the surface of the metallic rod. Now, when the rod makes physical contact with the sphere, the negatively charged electrons repel and spread out to the neutral sphere, thereby making the sphere negatively charged. The metal rod is still negatively charged, but it has less excess negative charge than it had prior to the conduction charging process. This process is called charging by *conduction*.

Electrostatic induction

Induction is charging an object without actually touching the object. Let us demonstrate this using a simple example. Consider an uncharged metallic sphere labeled *A* standing on an insulated stand so that the charge acquired by it is not transferred to the ground. We place another metallic sphere labeled *B* in contact with *A* so that the two spheres form a two-sphere system, as shown in figure 2.3(*a*). Since the spheres are conductors, the conduction electrons are free to move in the

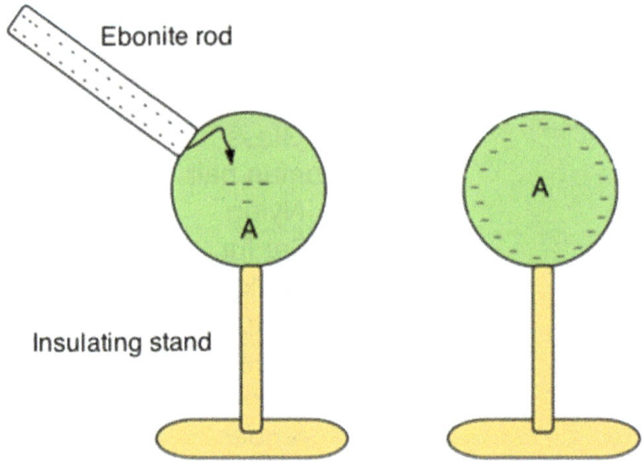

Figure 2.2. Charging a sphere by conduction.

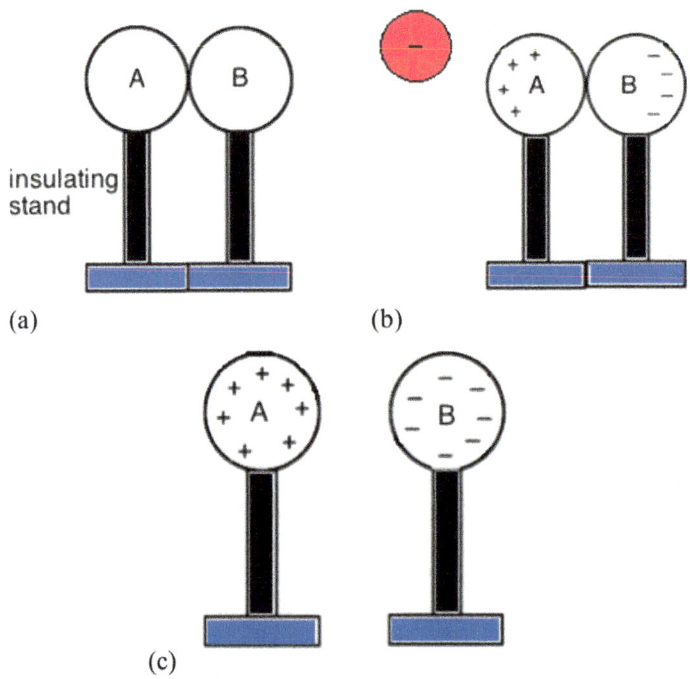

Figure 2.3. The process of induction.

two-sphere system. Now, a negatively charged sphere is brought near sphere *A*. The free electrons present in *A* are repelled and move to the farther end of *B*, as shown in figure 2.3(*b*). Overall the system is neutral. However, the mass-migration of electrons to system *B* creates an excess of electrons in *B* and a deficit of electrons in *A*. Looking at the spheres individually, this would mean that sphere *A* is positively charged and sphere *B* is negatively charged. Now, we separate the spheres by their insulating stands, as shown in figure 2.3(*c*). The excess negative charge remains on *B* and the deficit of negative charge remains with *A* and hence, we have *induced* equal and opposite charge on two spheres, *A* and *B*.

2.3 Some properties of charges

Conservation of charge

In all the electrification processes, we have seen that the total charge in an isolated system (no matter is allowed to enter the system from the outside) never changes. This is in fact true in a very general sense.

Consider an isolated empty system. We can let gamma rays pass into or out of the system. The high-energy photon may end its existence with the creation of an electron and a positron. Even though the two new particles are created, the total charge of the isolated system is still zero. This is the sense in which we mean the total charge of the system is conserved. A violation of the law of conservation of charge would be the creation of a positive charge without the simultaneous creation of an equal negative charge. Such an event has never been observed in practice. The theory of electromagnetism that we are going to develop in this book is strongly based on the law of conservation of charge. Hence, we may take this as an axiom.

Charge is neither created nor destroyed but is transferred from one body to another.

Quantization of charge

The electric charges we observe in the macroscopic world only come in units of one magnitude. This means that any amount of charge we observe can be written as an integral multiple of a number and that number is equal to the magnitude of the charge of the electron ($\sim 1.6 \times 10^{-19}$ C). Thus, the charge exists in terms of distinct packets but not in continuous form. This is called charge quantization. The description of charge quantization lies beyond the scope of classical electromagnetism and in our book we shall assume that the macroscopic charges are large enough for us to ignore the fine quantization of charge and treat the objects as if they contain continuous charge distribution.

Invariance of charge

The charge is a scalar quantity. This not only means that it does not require any direction to specify it completely, but it also means that its value does not change as it is viewed from different inertial frames of reference. Mass is another concept that is a scalar. However, there are some books that say that mass changes as it is viewed from different inertial frames of reference. This is incorrect. Mass is a tag that goes with the identity of the particle rather than its motion through space. So, when we say mass, it is the rest mass that we are talking about. We shall not, however, specify something as rest mass in future. In the same way, we talk about the invariance of the quantity called charge. A charge that has five coulombs in one inertial frame of reference has the same five coulombs in every inertial frame of reference. Hence, in this sense, charge is a scalar quantity.

2.4 Coulomb's law

Coulomb's law describes the interaction of charged particles at rest. In the 1780s a military engineer named Charles Coulomb discovered what Ampère later called 'electrostatics'. He posited two kinds of charges, positive and negative, and formulated the inverse square law by means of his torsion balance. He found that the electrical force between the two charges depended on their magnitude. He also observed that the force of attraction or repulsion decreases as $1/r^2$, where r represents the distance between the charged particles. Coulomb's law was crucial for the further development of electromagnetism. Let us try to understand Coulomb's law quantitatively.

Let us suppose that we have two charged particles q_1 and q_2 separated by a distance d. Let F be the magnitude of the force between them.

Then, from Coulomb's law we have that

$$F \propto \frac{q_1 q_2}{d^2}. \tag{2.1}$$

For reasons that will become clear later, the constant of proportionality is written as $\frac{1}{4\pi\varepsilon_0}$. Thus the proportionality shown in equation (2.1) can be written as

$$F = \frac{1}{4\pi\varepsilon_0}\frac{q_1 q_2}{d^2}. \tag{2.2}$$

The value of the constant can be determined from the experiments performed by Coulomb. It turns out that $\frac{1}{4\pi\varepsilon_0} = 9 \times 10^9 \frac{\text{Nm}^2}{\text{C}^2}$. This gives us that $\varepsilon_0 = 8.85 \times 10^{-12} \frac{\text{C}^2}{\text{Nm}^2}$. The force between two charges of one Coulomb separated by a distance of 1 m is approximately 9×10^9 N.

We have already stated that charge comes in two types: positive and negative. Let us now try to understand the reason for such a notation. Consider Coulomb's law. The force of attraction between unlike charges and the force of repulsion between like charges can be neatly summed up in one vector equation if we call one charge positive and the other charge negative.

The force between the charges of any kind is given by

$$\vec{F} = \frac{1}{4\pi\varepsilon_0}\frac{q_1 q_2}{|\vec{r}\,|^2}\hat{r}. \tag{2.3}$$

In the above equation, q_1 and q_2 are two charged point particles separated by a distance r and $\hat{r}$ is the unit vector. The vector $\hat{r}$ has a unit magnitude and is along the line joining q_1 and q_2. It is directed away from the center of the line joining the point charges q_1 and q_2. If one of the charges is of a different kind (one is positive and the other is negative), then the force is towards the center of the line joining the two charges and hence we have combined the two kinds of charges in a single vector equation: *Coulomb's law*.

Conclusion

In summary, we have seen that defining charge is as difficult as defining mass or any other fundamental property in science. However, we now have a basic understanding of its properties and the formula for interaction with other charges: Coulomb's law. This law serves as the basis for the further mathematical development of the ideas of electromagnetism, which are explored in the forthcoming chapters.

Exercises

MCQs

1) A fundamental particle or an elementary particle is a particle that is not made up of smaller particles. What are the fundamental particles of an atom?
   a) Quarks, gluons and electrons
   b) Protons, neutrons and electrons
   c) Protons, neutrons and electrons
   d) None of the above
2) The work done in moving a charge q along a circle in the presence of another charge Q is
   a) Zero
   b) Positive
   c) Negative
   d) Dependent on the radius of the circle
3) Two metallic spheres A and B of equal mass m are charged by the method of induction. In this process, suppose that A acquires positive charge and B acquires an equal amount of negative charge. Suppose that the masses of the spheres A and B after the process of induction are m_A and m_B, respectively. Which of the following is true?
   a) $m_A = m_B = m$
   b) $m_A < m_B = m$
   c) $m_A < m$ and $m_B > m$
   d) $m_A > m$ and $m_B < m$
4) Suppose that cotton is rubbed with amber to create frictional electrification. According to the triboelectric series, which of them will acquire a positive charge?
5) We have seen that a fundamental particle or an elementary particle is a particle that is not made up of smaller particles. Which of the following is a fundamental particle?
   a) Atom
   b) Meson
   c) Proton
   d) Neutrino

Problems

1) Suppose that the charge that flows through a circuit in time t and $t + \mathrm{d}t$ is given by

$$\mathrm{d}q = \mathrm{e}^{-\frac{t}{\tau}}\,\mathrm{d}t,$$

where τ is a constant. Find the total charge that flows through the circuit between $t = 0$ and $t = \tau$.
2) An electric field of magnitude E just prevents a water droplet of mass m from falling. Find the charge of the droplet.
3) When an object charge Q is divided into two objects, prove that the force of repulsion between the two will only be maximized if the total charge Q is divided equally between the two objects.
4) Suppose we have two charged point particles with charge $-q$ and $2q$ and a distance r between them. Where can we place a third charge Q so that the total force experienced by the charge Q is zero?
5) The protons in a nucleus are separated by a distance of 1 fermi (1 fermi = 10^{-15} m). Find the electric force between two protons.

References

[1] Feynman R 1967 *The Character of Physical Laws* (Cambridge, MA: MIT Press) p 31
[2] Bartlett D F and Ward B F L 1977 Is an electron's charge independent of its velocity *Phys. Rev.* D **16** 3453
[3] Loeb L B 1945 The basic mechanisms of static electrification *Science* **102** 573–6

Chapter 3

Electrostatics

3.1 Introduction and the need for the concept of fields

We saw in the previous chapter that Coulomb's law can only be used to describe the force experienced by charged particles when they are at rest with respect to each other. Coulomb's law is not precisely correct when the charges are moving. Of course, we can add terms to Coulomb's law to include the effects of the velocity and acceleration of charges. This idea of such a single equation for describing the interaction of charges was first proposed by Karl Friedrich Gauss in 1835. However, Weber published the first equation of this type in 1841. We shall now see the work of many brilliant scientists who have tried to develop electromagnetism along these lines.

The first step towards this direction is Coulomb's law. We have already seen that Coulomb's law states that the force **F** between two particles with charges Q_1 and Q_2 separated by a vector **r** can be written as

$$\vec{F} = \frac{1}{4\pi\varepsilon}\frac{Q_1Q_2}{r^3}\vec{r}.$$

The next step is to add an additional term to take into account the motion of charged particles. In conventional Maxwell theory, as we shall understand later, this is represented by a field called the *magnetic field.* However, another French scientist named Ampère postulated such a force between two current elements by experiments [1] and concluded that the force **F** between two current elements ds_1 and ds_2 carrying current I_1 and I_2 is given by

$$d^2\vec{F} = \vec{a}_r\frac{|I_1||I_2|\ ds_1ds_2}{4\pi\varepsilon c^2r^2}\Big[2\sin\theta_1\sin\theta_2\cos\eta - \cos\theta_1\cos\theta_2\Big],$$

where r is the distance between the current elements, the angle between ds_1 and $\mathbf{a}_r$ is θ_1 and the angle between ds_2 and $\mathbf{a}_r$ is called θ_2. The planes determined by $\mathbf{a}_r\ ds_1$ and $\mathbf{a}_r\ ds_2$ make the angle η. Ampère's law expresses the force between current elements

doi:10.1088/978-0-7503-1266-0ch3

and not between particles. Hence, it is not the desired final equation. Gauss in 1867 [2] modified Coulomb's law and Ampère's law and gave the following equation:

$$\vec{F} = \vec{a}_r \frac{Q_1 Q_2}{4\pi\varepsilon r^2}\left[1 + \left(\frac{v}{c}\right)^2 \left(1 - \frac{3}{2}\cos^2\theta\right)\right].$$

Here, v represents the magnitude of the relative velocity between Q_1 and Q_2 and θ is the angle between $\mathbf{v}$ and $\mathbf{a}_r$. Later, Weber [3], a student of Gauss, suggested that the force $\mathbf{F}$ is given by the following equation:

$$\vec{F} = \vec{a}_r \frac{Q_1 Q_2}{4\pi\varepsilon r^2}\left[\frac{1}{2}\left(\frac{v}{c}\right)^2 \left(\cos^2\theta\right)\right].$$

Gauss's equation is the only equation that agrees with Ampère's experiments and modern electrodynamics. It can be used to calculate the force between charged particles, whether these particles are stationary or in uniform relative motion. It can also be used to calculate the force between two conductors carrying direct current through them. However, it fails to give a correct description of alternating currents, as they involve the acceleration of charges. Later, Weber added a term to include the effects of acceleration, and it is remarkable because it is the first electrodynamics equation to include the acceleration term in it. The force $\mathbf{F}$ according to Weber [3] can be written as

$$\vec{F} = \vec{a}_r \frac{Q_1 Q_2}{4\pi\varepsilon r^2}\left[1 - \frac{1}{2}\left(\frac{v}{c}\right)^2 \left(\cos^2\theta\right) + \frac{r}{c^2}\left(\frac{\mathrm{d}v}{\mathrm{d}t}\right)\cos\psi\right].$$

This equation was derived from the basis of energy conservation and hence fails when applied to open circuit problems such as the radiation from antennae, etc. Hence, we see that it is not very easy to give a formula for the force produced by one charged particle on another. However, there is another view in which the laws of electrodynamics can be expressed in the simplest form. This is through the introduction of electric and magnetic fields. Michael Faraday first introduced the electric and magnetic fields in late 1855 as a 'constant necessary condition to action in space' [4]. We shall in subsequent chapters study how the introduction of electric and magnetic fields lets us express the laws of electromagnetism in a very simple way.

3.2 Electromagnetic fields

Suppose that we have a charged particle Q *fixed* at a point in space. At a certain distance r suppose that we have another charge q. Coulomb's law gives the force $\mathbf{F}$ experienced by the charge q as

$$\vec{F} = \frac{Qq}{4\pi\varepsilon_0 r^3}\vec{r}. \tag{3.1}$$

We can rewrite equation (3.1) as

$$\frac{\vec{F}}{q} = \frac{Q}{4\pi\varepsilon_0 r^3}\vec{r}.$$

Suppose we now bring in a new charge q_{new} at a position r from the charge. The force experienced by the new charge is given by

$$\vec{F}_{\text{new}} = \frac{Qq_{\text{new}}}{4\pi\varepsilon_0 r^2}\hat{r} = \frac{\vec{F}}{q}q_{\text{new}}$$

Hence, we can write the quantity $\dfrac{\vec{F}}{q}$ as **E** and we shall call it the electric field. Thus, if we have a charged particle at a fixed point in space, the electric field created by the charged particle at a position vector **r** is given by

$$\vec{E} = \frac{Q}{4\pi\varepsilon_0 r^3}\vec{r}.$$

This electric field is a function of the space and time coordinates. In a similar way, to include the effects of the motion of charged particles, we postulate the existence of another field called the magnetic field **B** (which is also a function of the space and time coordinates), so that the total force **F** experienced by a particle of charge q and moving with a velocity v with respect to these fields is

$$\vec{F} = q(\vec{E} + \vec{v} \times \vec{B}).$$

The electric and magnetic fields themselves are produced by the charged particles. Suppose that a certain configuration of charges produces an electric field $\mathbf{E}_1$ at a certain point, and another system of charges produces a field $\mathbf{E}_2$ at the same point. Then, the field $\mathbf{E}_{\text{new}}$ produced when the two systems of charges are present simultaneously is given by

$$\vec{E}_{\text{new}} = \vec{E}_1 + \vec{E}_2.$$

The same is also true for magnetic fields. This fact is called the *principle of the superposition of fields*. This is an extremely important property of electric and magnetic fields.

We can try to obtain a mental picture of the electric and magnetic fields by drawing vectors at many different points in space. Such a representation is shown in figure 3.1.

The lengths of the vectors in the diagram represent the magnitude of the field at those points in space. We can also draw lines that are everywhere tangent to these vectors. In this case, we lose track of the length of the vectors, but the magnitude of the field at any given point is proportional to the number of field lines at that point. This is shown in figure 3.2.

3.3 The concept of flux

Consider an electric field **E** present in space and consider area element d**S** whose normal forms an angle θ with the field lines as shown in figure 3.3.

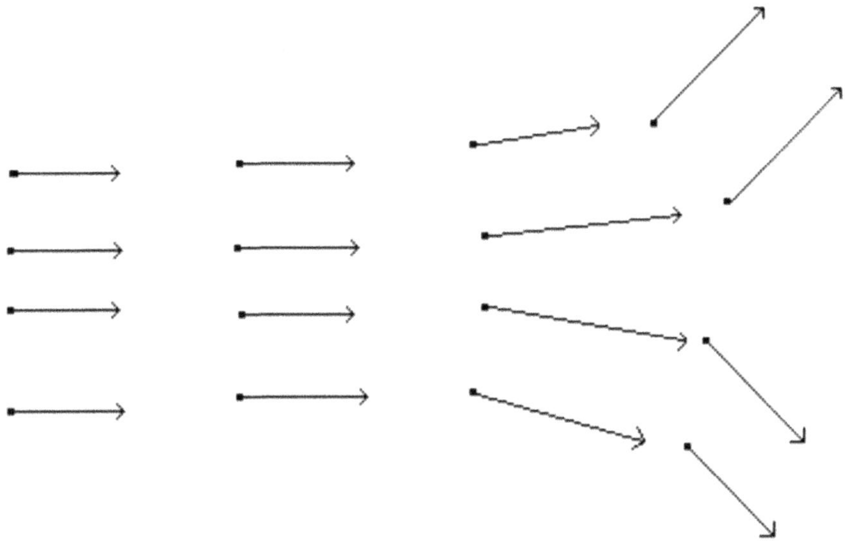

Figure 3.1. A vector field represented by arrows at different points in space.

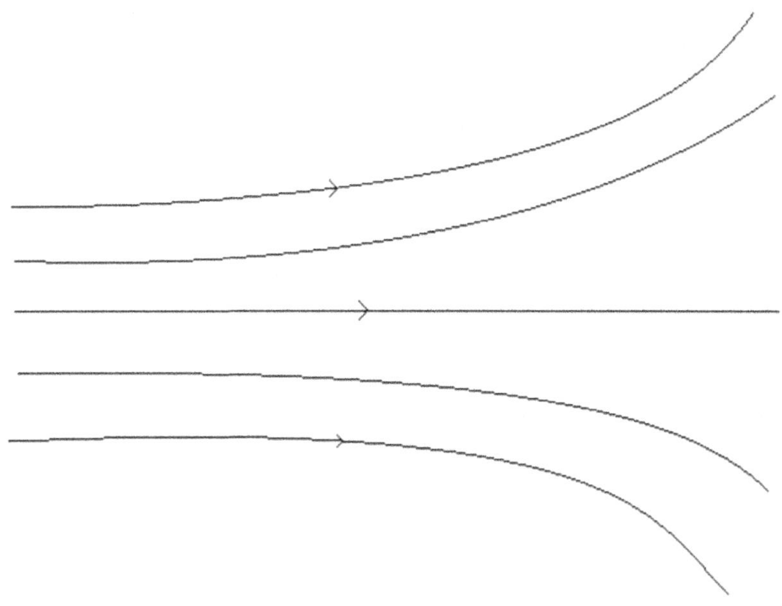

Figure 3.2. Field lines showing a vector field in space.

The number of lines piercing the area element is given by dΦ. From the diagram shown in figure 3.3, we can write that

$$d\Phi = E \times dS \cos\theta = \vec{E} \cdot d\vec{S}.$$

This quantity is called the flux of electric field through the area element **dS**. It should be noted that the choice of the direction of the area vector is arbitrary. This vector

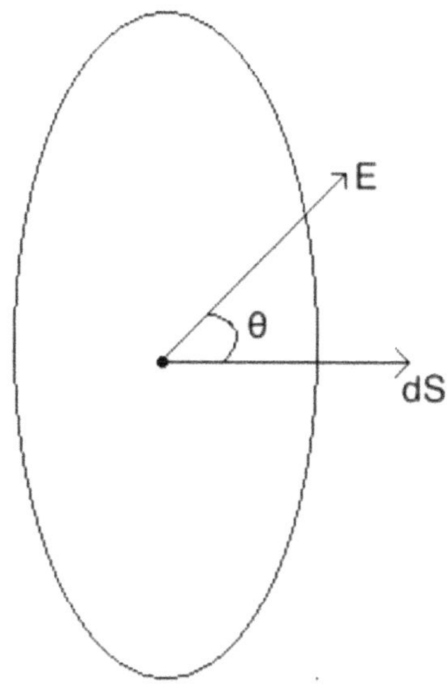

Figure 3.3. An area element that makes an angle θ with the electric field present in space.

could be directed oppositely. But, if we have a closed surface, the area vector at all points on the closed surface can be taken to be directed outside the region enveloped by the surface. The flux of the electric field through this closed surface is then given by

$$\Phi = \oint_S \vec{E} \cdot \mathrm{d}\vec{S}.$$

Even though we have only used electric field to define the flux, we can use any vector and the concept of flux is applicable in the same way.

3.4 Gauss's theorem

The flux of the electric field **E** through a closed surface **S** has a remarkable property in that it only depends on the total amount of charge present in the closed surface. This relationship is called Gauss's law and is written as

$$\Phi = \oint_S \vec{E} \cdot \mathrm{d}\vec{S} = \frac{q_{\text{in}}}{\varepsilon_0}.$$

This is called the *Gauss's theorem*: the flux of **E** through a closed surface **S** is equal to the algebraic sum of the charges enclosed by this surface, divided by ε_0. Let us see this in a simple example. Consider a point charge Q and a sphere with a radius r surrounding it. This sphere is called a Gaussian surface. The electric field created by the point charge at a distance r is given by

$$\vec{E} = \frac{1}{4\pi\varepsilon_0}\frac{Q}{r^3}\hat{r}.$$

We shall now divide the Gaussian surface into small patches with area **dS** and directed outward. The flux is then given by

$$\Phi = \oint_S \vec{E} \cdot \mathrm{d}\vec{S}.$$

The electric field at every point on the Gaussian surface is in the same direction as **dS** and is constant in magnitude. Hence, we can write that

$$\Phi = \frac{1}{4\pi\varepsilon_0}\frac{Q}{r^2}\oint_S \mathrm{d}\vec{S}.$$

The integration of **dS** on the Gaussian surface gives us the surface area. Hence, we have that

$$\Phi = \frac{1}{4\pi\varepsilon_0}\frac{Q}{r^2}4\pi r^2.$$

Finally, it gives us that

$$\Phi = \oint_S \vec{E} \cdot \mathrm{d}\vec{S} = \frac{Q}{\varepsilon_0}.$$

Hence, we have proved Gauss's theorem in the case of a single point charged particle. It can be proved that if the charge q lies outside the Gaussian surface, its flux through the surface is zero. In terms of field lines this means that the number of lines entering the closed surface is equal to the number of lines leaving the closed surface. We can now generalize the case where the electric field is produced by a system of point charges represented as Q_1, Q_2, Q_3, ..., Q_n. The electric field created by them is represented as $\mathbf{E}_1$, $\mathbf{E}_2$, $\mathbf{E}_3$, ..., $\mathbf{E}_n$. The total flux of the electric field created by the system of charges is

$$\Phi = \oint_S \vec{E} \cdot \mathrm{d}\vec{S} = \oint_S \left(\vec{E}_1 + \vec{E}_2 + \vec{E}_3 + \cdots + \vec{E}_n\right) \cdot \mathrm{d}\vec{S}$$

$$\Rightarrow \Phi = \oint_S \vec{E} \cdot \mathrm{d}\vec{S} = \oint_S \vec{E}_1 \cdot \mathrm{d}\vec{S} + \oint_S \vec{E}_2 \cdot \mathrm{d}\vec{S} + \oint_S \vec{E}_3 \cdot \mathrm{d}\vec{S} + \cdots + \oint_S \vec{E}_n \cdot \mathrm{d}\vec{S}. \tag{3.2}$$

In accordance with the previously mentioned statement, each integral on the right side of equation (3.2) is equal to q_i/ε_0 if q_i is present in the closed surface. Otherwise, it is zero. Thus, the right side of equation (3.2) will only contain the algebraic sum of those charges that lie inside the closed surface **S**. Now, the flux of the electric field through a closed surface will remain constant as long as the charges inside the surface do not go outside the surface. This means that if we displace the charges, the field created by them will change everywhere, but the flux of this electric field through the surface will remain unchanged. This is a remarkable property of the flux of an electric field.

3.5 Differential form of the Gauss theorem

In the previous section, we only considered the case where the charges are discrete point particles in space. Let us consider the case where the charges are distributed continuously with the volume density ρ, depending on the Cartesian coordinates. The charge present in a volume V can then be written as

$$q_{\text{in}} = \int \rho \cdot \mathrm{d}V.$$

Let us represent the charge q in the volume V enclosed by a closed surface S as

$$q_{\text{in}} = \rho_{\text{avg}} V$$

Here, ρ_{avg} represents the volume charge density, averaged over the volume V. Using Gauss's theorem for this closed surface S, we get that

$$\frac{1}{V} \oint_S \vec{E} \cdot \mathrm{d}\vec{S} = \frac{\rho_{\text{avg}}}{\varepsilon_0}.$$

We now make the volume V tend to zero by contracting it to a point P. This means that the ρ_{avg} becomes the charge density ρ at point P. Hence, we get that

$$\lim_{V \to 0} \frac{1}{V} \oint_S \vec{E} \cdot \mathrm{d}\vec{S} = \frac{\rho}{\varepsilon_0}. \tag{3.3}$$

We saw in chapter 1 that the left-hand side of equation (3.3) is just the divergence of the electric field at the particular point P. Hence, we have that

$$\nabla \cdot \vec{E} = \frac{\rho}{\varepsilon_0}.$$

This is the differential form of Gauss's theorem and it is also, as we shall see later, Maxwell's first equation.

Conclusion

In electromagnetism, the idea of field was first introduced by Michael Faraday to explain his experimental results. This has been by far the most efficient way of describing electricity and magnetism. We have seen in this chapter that adding terms to include the effects of velocity and acceleration to explain the forces between the two charged particles results in a complex description of electromagnetism. The idea of field introduced in this chapter is crucial to further development of the theory. We have also learnt about the concept of electric flux and its relationship to charges.

Exercises

MCQs

1) Figure M3.1 shows some of the electric field lines corresponding to an electric field with magnitude E. What does the figure suggest?

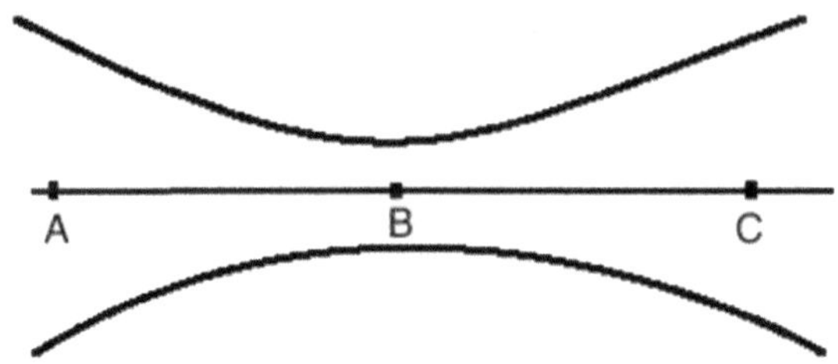

Figure M3.1.

a) $E_1 > E_2 > E_3$
b) $E_1 = E_3 < E_2$
c) $E_1 = E_2 = E_3$
d) $E_1 = E_3 > E_2$

2) Suppose that a proton and an electron are placed in a uniform electric field. Which of the following is true?
   a) The electric forces acting on them are equal
   b) Their acceleration will be equal
   c) The magnitude of the electric forces on them is equal
   d) The magnitude of the acceleration will be equal
3) The total force **F** experienced by a charged particle of charge q moving with a velocity **v** in an electric field **E** and a magnetic field **B** is given by:
   a) $\vec{F} = q(\vec{E} + \vec{B})$
   b) $\vec{F} = q(\vec{B} + \vec{v} \times \vec{E})$
   c) $\vec{F} = q(\vec{v} \times \vec{B} + \vec{E})$
   d) None of the above
4) A metallic particle with no net charge is placed near a finite metal plate carrying a negative charge. The electric force experienced by the particle is:
   a) Towards the plate
   b) Away from the plate
   c) Zero
   d) None of the above
5) A positive point charge q is brought near an isolated metal cube. Which of the following is true?
   a) The cube becomes positively charged
   b) The cube becomes negatively charged
   c) The interior remains charge-free but the surface obtains a non-uniform charge distribution
   d) None of the above

Problems

1) A uniform electric field of magnitude **E** exists in space in the x-direction. Calculate the flux of this field through a plane square of edge a placed in the yz plane.

2) The intensity of an electric field depends on the x and y coordinates as shown:

$$\vec{E} = \frac{k\left(x\hat{i} + y\hat{j}\right)}{\left(x^2 + y^2\right)}.$$

Here, k is a constant. Find the charge within a sphere of radius r with its center at its origin.

3) A uniform electric field exists in space in the x-direction. Find the flux of this field through a cylindrical surface with the axis parallel to the x-axis.
4) A charge Q is uniformly distributed on a ring of radius R. A Gaussian sphere of radius R is formed with its center at the periphery of the ring, as shown in figure M3.2. Find out the flux of the electric field through the surface of the sphere.

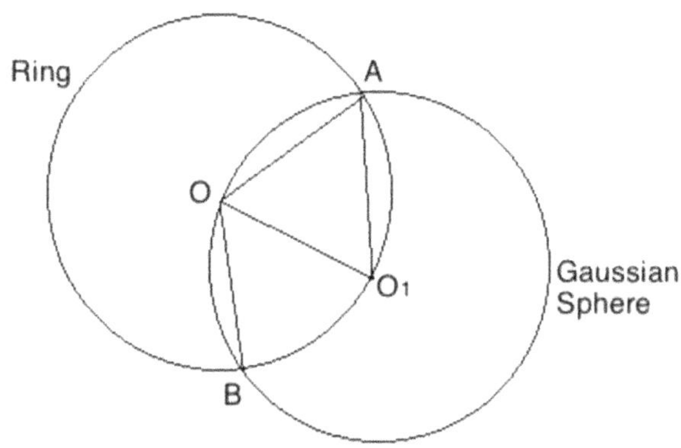

Figure M3.2.

References

[1] Ampère A-M 1823 *Mèm. Acad. Sci.* **6** 175–388

[2] Gauss C F 1867 Zur mathematischen Theorie der elektrodynamischen Wirkung *Werke* vol 5 Gottingen p 602

[3] Weber W 1893 Eletrodynamische Maassbestimmungen uber ein allgemeines Grundgesetz der elektrischen Wirkung *Werke* vol 3 (Berlin: Springer) p 25

[4] Roche J 2016 Introducing electric fields *Phys. Educ.* **51** 055005

IOP Publishing

The Foundations of Electric Circuit Theory

N R Sree Harsha, Anupama Prakash and D P Kothari

Chapter 4

The electric potential

In order to move a mass upward in the Earth's gravitational field, some external force is required. Work done by the external force increases the potential energy of the mass. By contrast, movement of the mass in the downward direction in the Earth's gravitational field does not require any external force and, thus, there is a decrease in the potential energy of the mass.

Similarly, movement of a positive test charge within an electric field changes its potential energy. On comparing the positive charge with mass and electric field with Earth's gravitational field, one can say that the movement of the positive charge against the electric field increases its potential energy and the movement of the positive charge in the direction of the electric field decreases its potential energy; the opposite is true for a negative charge.

4.1 The electric potential difference

The electric potential helps in understanding the electric phenomenon and is termed the potential energy per unit charge. The work done in moving a unit charge from one point to another within an electric field is equal to the difference in the potential energies at each point.

$$\text{Change in electrical potential energy } (\Delta U) = \text{work done } (W),$$

$$\text{electric potential} = \frac{(W)\ or\ (\Delta U)}{q_{\text{moved}}}.$$

Let us move a positive charge q from point A to point B within an electric field. The potential energy of the charge q at points A and B is U_A and U_B, respectively. *Since the electrostatic force is conservative in nature, the work done by it does not depend on the path followed by the particle but depends on the initial and final position of the particle*. If the work W is positive, i.e. the force and displacement are in the same direction, the potential energy P_B is smaller than P_A. If the work W is negative,

doi:10.1088/978-0-7503-1266-0ch4

i.e. the force and displacement are in the opposite direction, the potential energy P_B is greater than P_A.

$$W = (U_A - U_B) = -(U_B - U_A).$$

Since work = force × distance,

$$U_B - U_A = -W = -\int_A^B \overline{F} \cdot \mathrm{d}\overline{L}$$

$$V_B - V_A = \frac{U_B - U_A}{q} = -\frac{1}{q}\int_A^B \overline{F} \cdot \mathrm{d}\overline{L}.$$

In practice, it is convenient to consider the reference point to be at infinity ($A \to \infty$) with zero potential, and then the electric potential energy at point B is

$$U_B = -\int_\infty^B \overline{F} \cdot \mathrm{d}\overline{L}.$$

The potential V_B is given as

$$V_B = \frac{U_B}{q} = -\frac{1}{q}\int_\infty^B \overline{F} \cdot \mathrm{d}\overline{L}. \tag{4.1}$$

Since the force per unit charge is the electric field, equation (4.1) is rewritten as

$$V_B = -\int_\infty^B \overline{E} \cdot \mathrm{d}\overline{L}. \tag{4.2}$$

Since the unit of work or energy is the joule (J) and the unit of charge is the coulomb (C), the electrical potential can be measured in joules/coulombs (J/C) and defined as a volt (V).

$$1\frac{\mathrm{J}}{\mathrm{C}} = 1\mathrm{V}.$$

Some important points to remember:

- Electric potential energy and electric potential are two different things.
- Electric potential is a scalar quantity.
- $1\mathrm{e} = 1.6 \times 10^{-19}$ C, thus 1 eV = 1.6×10^{-19} J. The eV is therefore not a smaller unit of the volt, it is a smaller unit of the joule

Let us consider two positive charge q_1 and q_2, where q_1 is stationary and q_2 is moved from point A at infinity to point C, as shown in figure 4.1. The magnitude of the electric field at a distance r due to charge q_1 is given by equation (4.3):

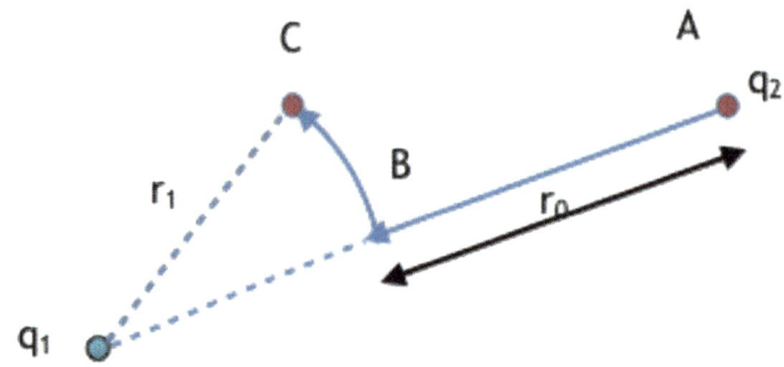

Figure 4.1. The charged particle q_2 is brought from infinity (A) to point C.

$$E = \frac{q_1}{4\pi\varepsilon_0 r^2}. \tag{4.3}$$

The electrostatic potential at point C can be evaluated by integration using equation (4.2) along the path. The path BC is perpendicular to the electric field produced by charge q_1 and therefore there is no change in the electrostatic potential from point B to point C. Thus the initial position of q_2 is at infinity and the final position is at distance r_1 from q_1.The electrostatic potential is

$$V_{r_1} = -\int_{\infty}^{r_1} \overline{E} \cdot d\overline{r}. \tag{4.4}$$

Substituting the value of the electric field from equation (4.3) into equation (4.4),

$$V_{r_1} = \int_{\infty}^{r_1} \frac{q_1}{4\pi\varepsilon_0 r^2} dr = \frac{q_1}{4\pi\varepsilon_0 r_1}.$$

Thus for a positive charge q_1, the potential decreases with distance and the direction of the electric field is from higher potential to lower potential. The potential energy of the charge q_2 under the influence of the electric field produced by charge q_1 is

$$U(r) = q_2 V(r) = \frac{1}{4\pi\varepsilon_0}\frac{q_1 q_2}{r}.$$

4.2 Earnshaw's theorem

In 1842, Samuel Earnshaw stated that it is not possible to have a stable array of particles in static equilibrium solely through the forces, which are governed by the inverse square law. The inverse square law governs all the gravitational, electrostatic and magnetic forces. The implications this theorem had for Victorian ideas regarding the constitution of matter are described in [1].

In order to understand Earnshaw's theorem, let us consider a simple system of two particles with charges q_1 and q_2 at a distance $\mathbf{r}$ between them. Let the potential energy of this system be U. Now suppose that the distance shrinks by a factor of λ, so that the new distance between the particles is $\mathbf{r}_1$. Hence, we have that

$$\bar{r}_1 = \frac{\bar{r}}{\lambda}.$$

The new potential energy U_1 will be a function of λ. Now, if our system of two particles is to be stationary, the potential energy function should be independent of changes to the scale of the distance between them. In particular, we must have that

$$\frac{dU_1}{d\lambda} = 0.$$

For stable equilibrium, we must also have that

$$\frac{d^2U_1}{d\lambda^2} > 0.$$

However, from the potential energy function defined in the previous section, we have that

$$U_1 = \frac{1}{4\pi\varepsilon_0}\frac{q_1q_2}{\left(\frac{r}{\lambda}\right)} = \frac{\lambda}{4\pi\varepsilon_0}\frac{q_1q_2}{r} = \lambda U.$$

Hence, we have that

$$\frac{dU_1}{d\lambda} = U.$$

And, we also have that

$$\frac{d^2U_1}{d\lambda^2} = 0.$$

Hence, we can only have a stable equilibrium for this system of two particles if $U = 0$. Let us also try to understand this with the help of Gauss's theorem. Consider the diagram shown in figure 4.2.

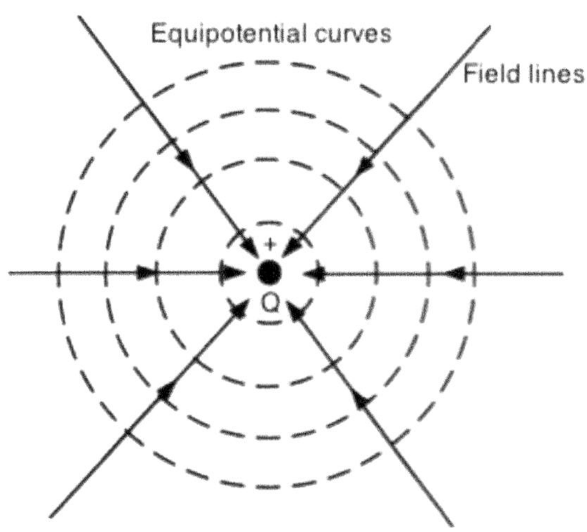

Figure 4.2. An impossible situation is shown here. The field lines are represented by solid lines and the equipotential curves are represented by dashed lines.

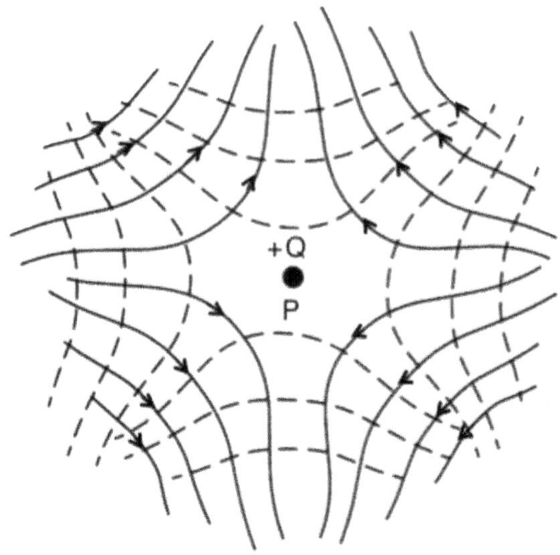

Figure 4.3. The charge Q is in unstable equilibrium if it is placed in the saddle point P.

If a positive charge Q is to be in stable equilibrium, it must be at the minimum of the electrostatic potential. The field lines should all point inwards, as shown in figure 4.2. Now, apply Gauss's law to a spherical surface on one of the equipotential curves. The field lines should all point inwards, which means that the divergence of this electric field is negative. But, according to Gauss's law, this can only happen if there is a negative charge inside the surface. Since Q is positive, we have proved that it is not possible to create the field lines as shown in figure 4.2. Hence, a positive charge Q can never be in a stable equilibrium under the influence of electric forces alone. However, the charge Q can be in an unstable equilibrium under the influence of electric forces. One example of this is to place the charge Q in a saddle point P, as shown in figure 4.3.

4.3 Conductors and insulators

A conductor is a material in which electrons are loosely bound to the nucleus and are free to flow in the material. Any external influence that moves one of them causes others to repel and propagate this effect through the conductor.

When the conductor is subjected to a potential difference, this results in the movement of negative charges towards the positive end of the potential difference. This flow of charges is called electric current.

The capacity to carry current, i.e. the ampacity of a conductor, is related to its resistance. Lower resistance will result in higher flow of current through a conductor and vice versa. The resistance of a conductor depends upon its material and dimensions. The resistance R of a conductor of uniform cross-section is given by equation (4.5):

$$R = \rho \frac{l}{A}, \tag{4.5}$$

where ρ is the electrical resistivity of the conductor's material in $\Omega\ m^{-1}$, l is the length of the conductor in meters and A is the area of the cross-section of the conductor in square meters. Conductors are common metals like aluminum, copper, etc.

An insulator is a material in which even the outermost electrons are tightly bound to the nucleus and thus cannot conduct current when subjected to potential difference. There is no perfect insulator with infinite resistivity, however materials like glass, mica and paper have high resistivity and can serve effectively as insulators. Insulators, when subjected to a very high voltage, experience an electrical breakdown, in which current suddenly rises through the material, leading to the thermal breakdown of the material.

4.4 Capacitors

A capacitor is a passive electrical element that stores electrical energy temporarily in the electrostatic field. A basic capacitor is made up of two conducting plates separated by either air or some insulating material, such as waxed paper, mica, ceramic or plastic, known as a dielectric.

Capacitance is an electrical property of a capacitor. For the basic and commonly used parallel plate capacitor, the capacitance is dependent upon the area A (square meters) of the smaller of the two plates, the distance d (meters) between the two plates and the absolute permittivity ϵ of the dielectric used between the two plates, and is given by equation (4.6):

$$C = \frac{\epsilon A}{d}, \tag{4.6}$$

where $\epsilon = \epsilon_0 \epsilon_r$, ϵ_0 is the permittivity of free space, a constant with the value $8.84 \times 10^{-12}\ F\ m^{-1}$, and ϵ_r is the relative permittivity of the dielectric material.

Capacitance which is the charge storing property of a capacitor is defined as the ratio of the electric charge q on the conductor to the potential difference V between them and is measured in farads (F).

$$C = \frac{q}{V}\ \text{(farad)}. \tag{4.7}$$

One farad means precisely 6.28×10^{18} stored electrons, which is a huge number and thus capacitors are rated in either microfarads (10^{-6}) or picofarads (10^{-12}).

When a capacitor is connected across a battery, an electric field is developed across the dielectric, resulting in the accumulation of positive charges on one plate and negative charges on the opposite plate. This accumulation of charges develops a potential difference across the two plates. The flow of electrons onto the plates continues until the potential difference across the two plates is equal to the potential difference of the battery connected across it. At this point the capacitor is fully charged with the maximum electrostatic field. Likewise, when the charged capacitor is connected across a load, it discharges and the potential difference across the capacitor plates decreases with the decrease in electrostatic field as the energy moves out from the capacitor to the load.

Relationship between the voltage and current of a capacitor
The relationship between the voltage and current of a capacitor defines its capacitance and power. In order to derive this relationship, let us rewrite equation (4.7)

$$q(t) = C\,v(t).$$

On taking a derivative of it, we obtain

$$\frac{\mathrm{d}q(t)}{\mathrm{d}t} = C\frac{\mathrm{d}v(t)}{\mathrm{d}t}. \tag{4.8}$$

Since $\frac{\mathrm{d}q(t)}{\mathrm{d}t} = i(t)$, the current flowing through the capacitor,

$$i(t) = C\frac{\mathrm{d}v(t)}{\mathrm{d}t}. \tag{4.9}$$

On analyzing equation (4.9), it can be concluded that the current flows through a capacitor only when it is subjected to a varying voltage. Thus, for a constant battery source, the capacitors act as an open circuit, because no current flows through it.

The voltage across the capacitor can be written in terms of current flowing through it by integrating equation (4.9), as shown by equation (4.10):

$$v(t) = \frac{1}{c}\int_0^t i(t)\mathrm{d}t + v_c(0), \tag{4.10}$$

where $v_c(0)$ is the voltage across the capacitor at $t = 0$ (the initial voltage across the capacitor).

Behavior of a capacitor in a DC circuit
In order to understand the behavior of a capacitor in a DC circuit, let us consider a DC circuit consisting of resistance R, capacitance C, a switch S and a constant DC source of voltage V_S as shown in figure 4.4.

If the switch is closed at t_0, with zero initial charge at the capacitor, then as per the Kirchhoff voltage law (KVL), equation (4.10) gives the voltage equation for the DC circuit.

$$\begin{aligned} V_S &= v_r(t) + v_c(t) \\ V_S &= i(t)R + \frac{1}{C}\int_{t_0}^t i(t)\mathrm{d}(t). \end{aligned} \tag{4.11}$$

On rewriting equation (4.11) by taking the derivative of it, we obtain

$$\mathrm{RC}\frac{\mathrm{d}i(t)}{\mathrm{d}t} + i(t) = 0. \tag{4.12}$$

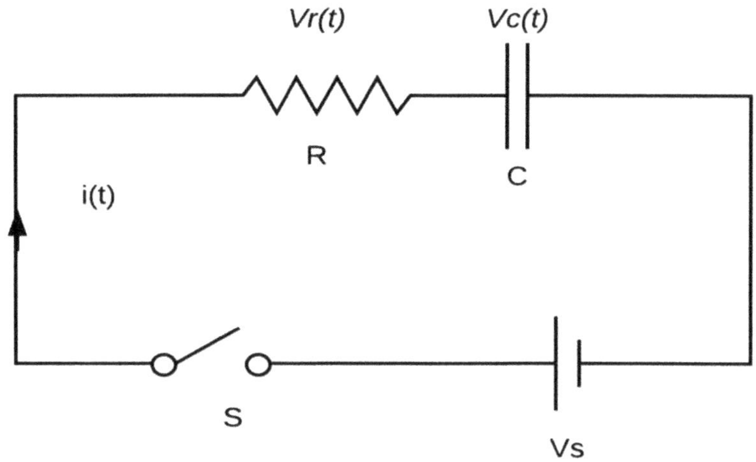

Figure 4.4. Resistor–capacitor (RC) circuit with a DC source.

Since equation (4.12) is a first-order differential equation, the solution to this homogeneous equation is

$$i(t) = Ae^{-\frac{1}{RC}t}. \tag{4.13}$$

At $t = 0$, the voltage across the capacitor is zero and it acts as a short circuit, resulting in $i_0 = V_S/R$. On putting this condition into equation (4.13), we obtain $A = V_S/R$.

Thus the complete solution of equation (4.13) is

$$i(t) = \frac{V_S}{R}e^{-\frac{t}{\tau_0}} \tag{4.14}$$

$$v_c(t) = V_S\left(1 - e^{-\frac{t}{\tau_0}}\right), \tag{4.15}$$

where $\tau_0 = RC$ is the time constant of the system.

It is evident from equations (4.14) and (4.15) that while charging current flowing through a capacitor decreases exponentially, the voltage increases exponentially, as shown in figure 4.5.

In the steady state condition, when the capacitor is fully charged, i.e. $V_c = V_S$, if the voltage source is disconnected and the two terminals are short-circuited, then the capacitor starts discharging. This process continues until the voltage across the capacitor becomes zero.

On applying KVL to the circuit shown in figure 4.4, we obtain

$$\begin{aligned} v_r(t) + v_c(t) &= 0 \\ Ri(t) + v_c(t) &= 0 \\ RC\frac{dv_c(t)}{dt} + v_c(t) &= 0. \end{aligned} \tag{4.16}$$

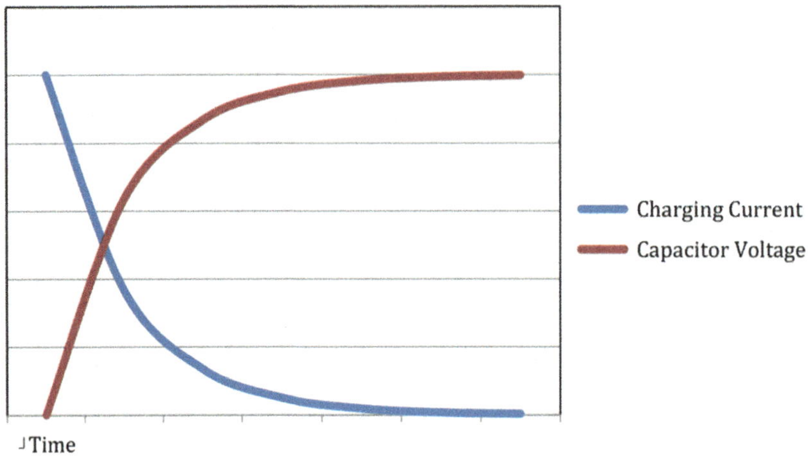

Figure 4.5. Graph of the charging current and voltage.

The solution of the homogeneous equation (4.16) is

$$v_c(t) = B\mathrm{e}^{-\frac{t}{RC}} \tag{4.17}$$

At $t = 0$, $v_c(t) = V_S$, and on putting this condition into equation (4.17), we obtain $B = V_S$.

Thus the complete solution of equation (4.17) is

$$v_c(t) = V_S\mathrm{e}^{-\frac{t}{RC}}. \tag{4.18}$$

Similarly,

$$i_c(t) = \frac{V_S}{R}\mathrm{e}^{-\frac{t}{RC}}. \tag{4.19}$$

(Note: the flow of the current will be in the opposite direction to that of the current during charging.)

From equations (4.18) and (4.19), one can conclude that both the voltage and current decrease exponentially, as shown in figure 4.6.

Behavior of a capacitor in an AC circuit

When a capacitor is connected across an AC voltage source, as shown in figure 4.7, then as per the KVL, the voltage equation is

$$V_S \sin \omega t - v_c(t) = 0$$

$$V_S \sin \omega t - \frac{q(t)}{C} = 0$$

$$q(t) = V_s C \sin \omega t.$$

Since $i = \mathrm{d}q(t)/\mathrm{d}t$,

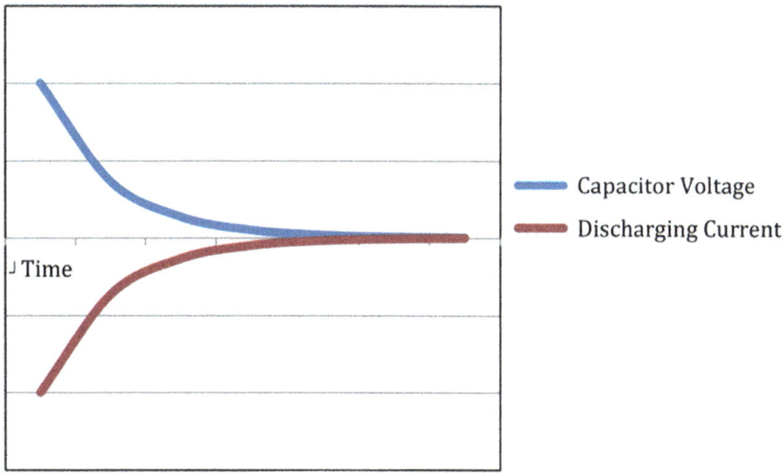

Figure 4.6. Graph of discharging current and voltage.

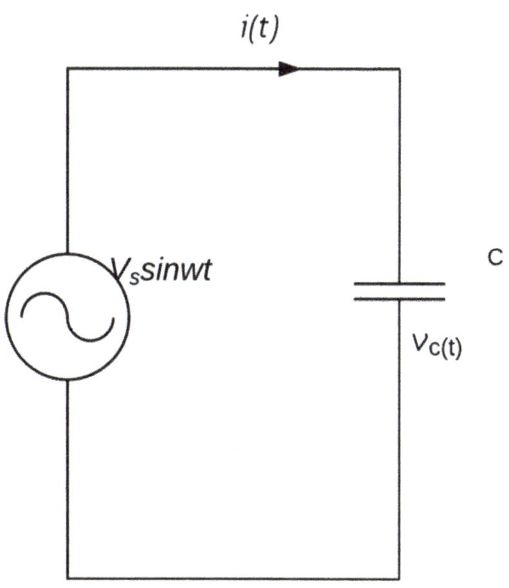

Figure 4.7. Capacitor connected across an AC supply.

$$
\begin{aligned}
i(t) &= \frac{\mathrm{d}(V_S C \sin \omega t)}{\mathrm{d}t} \\
i(t) &= V_S C\omega \cos \omega t \\
i(t) &= \frac{V_S \cos \omega t}{1/\omega C}.
\end{aligned}
\tag{4.20}
$$

As we know, the current through any element is equal to the voltage across the element divided by the resistive nature of the element. In equation (4.20), the numerator represents the voltage across the capacitor and the denominator, i.e., $1/\omega C$, is the resistive nature of the capacitor. This frequency-dependent resistive nature of the capacitor is termed the capacitive reactance, X_c.

$$X_c = \frac{1}{\omega C}\,(\text{ohms}). \tag{4.21}$$

As seen from equation (4.21), at high frequencies X_c approaches 0, i.e., the capacitor resembles a short wire that passes high-frequency signal. At low frequencies, X_c approaches infinity, i.e. the capacitor tends to behave as an open circuit that does not pass low-frequency signal.

4.5 The energy stored in a capacitor

The charging of a capacitor requires some amount of work to be done. The energy associated with this work is stored as electrical potential energy in the electrostatic field created between the two plates of a parallel plate capacitor.

Let us consider a capacitor with capacitance C connected across a DC voltage source that has voltage V with a switch S. Initially the capacitor does not have any charge or potential, i.e. $q = 0$, $V_c = 0$. At the time of switching, the full battery voltage is applied across the capacitor. A positive charge from the battery moves to the plate A of the capacitor, but during this movement no work is done by the battery. This first charge creates a small potential difference across the capacitor, and then the second positive charge is repelled by the first charge while moving to the positive plate of the capacitor. Since the battery voltage is more than the capacitor voltage, the second charge is also stored on the positive plate of the capacitor. Once some charge is transferred, an electric field is set up between the plates, which opposes any further charge transfer. In order to fully charge the capacitor, work is done against this field. This work becomes energy stored in the capacitor. Thus one can conclude that with increasing storage of charge, the voltage and energy of the capacitor increases.

The work done in transferring a small amount of charge $\mathrm{d}q$ is

$$\mathrm{d}W = V\mathrm{d}q.$$

The total work W done in transferring the total charge Q from one plate to the other is

$$W = \int_0^Q V\mathrm{d}q.$$

Since $V = q/c$,

$$W = \int_0^Q \frac{q\,\mathrm{d}q}{C} = \frac{Q^2}{2C} = \frac{CV^2}{2} = \frac{1}{2}QV. \tag{4.22}$$

From equation (4.22), the total amount of energy given by the battery is $W_{\text{total}} = QV$. This shows that half of this energy is stored in the capacitor and the remaining half is lost in the process.

If the capacitor considered is a vacuum-filled parallel plate capacitor with plates of cross-sectional area A, spaced at a distance d apart, then the electric field E between the plates is

$$E = \frac{Q}{A\epsilon_o} = \frac{\sigma}{\epsilon_0},$$

where $\sigma = \frac{Q}{A}$ is the charge density and ϵ_0 is the permittivity of the vacuum.

The potential difference between the plates is given as $V = Ed$. The energy stored in the capacitor can also be written as

$$W = \frac{CV^2}{2} = \frac{\epsilon_0 A\, E^2 d^2}{2d} = \frac{\epsilon_0 E^2 Ad}{2}.$$

Ad is the volume of the region between the plates, where energy is stored. Thus the energy per unit volume w is

$$w = \frac{\epsilon_0 E^2}{2}.$$

Conclusion

In this chapter, we have seen the definition of electrical potential energy and the idea of electric potential at a point in space. We have also seen Earnshaw's theorem. We have defined how conductors and insulators behave when external electric field is applied. We have studied the behavior of a capacitor when it is connected across different voltage sources.

Exercises

MCQs

1) When the separation between charges is decreased, the electric potential energy of the charges:
    a) Increases
    b) Decreases
    c) Remains the same
    d) May increase or decrease
2) If a negative charge is shifted from a low potential region to a high potential region, the electric potential energy:
    e) Increases
    f) Decreases
    g) Remains the same
    h) May increase or decrease
3) The magnitude of the electric field and the electric potential at a point in space are E and V, respectively. Which of the following is true?
    a) If $E = 0$, V must be zero
    b) If $V = 0$, E must be zero

c) If $E \neq 0$, V cannot be zero
d) None of the above

4) Which of the following quantities depend on the choice of zero potential?
   a) Potential at a point
   b) Potential difference between two points
   c) Electric field at a point
   d) None of the above
5) A capacitor of capacitance C is charged to a potential V. Suppose we have a closed surface S that encloses the capacitor. The flux of electric field through the surface S multiplied by ϵ_0 is equal to:
   a) CV
   b) $2CV$
   c) $CV/2$
   d) Zero

Problems

1) Two charges 10 nC and 20 nC are placed at a distance of 2 m. Find the electric potential due to the pair at the middle of the line joining the two charges.
2) Three charged particles that each have a charge of 10 nC are placed at the vertices of an equilateral triangle with a side of 10 cm. Find the work done by an external agent in pulling them apart to infinite separation.
3) Find the equivalent capacitance between the points A and B in the diagram shown in figure M4.1.

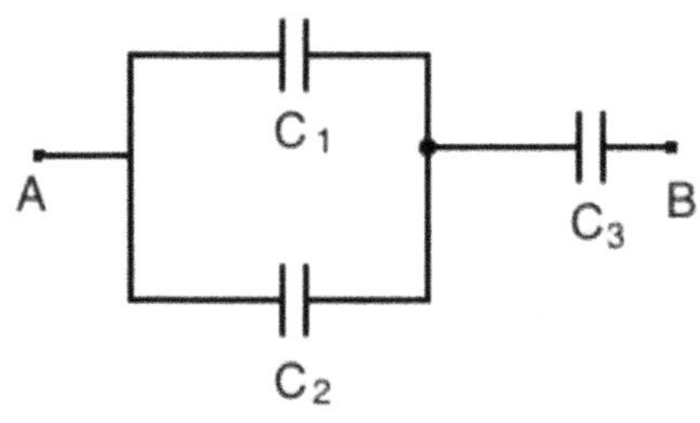

Figure M4.1.

4) Calculate the charge on each capacitor shown in figure M4.2.

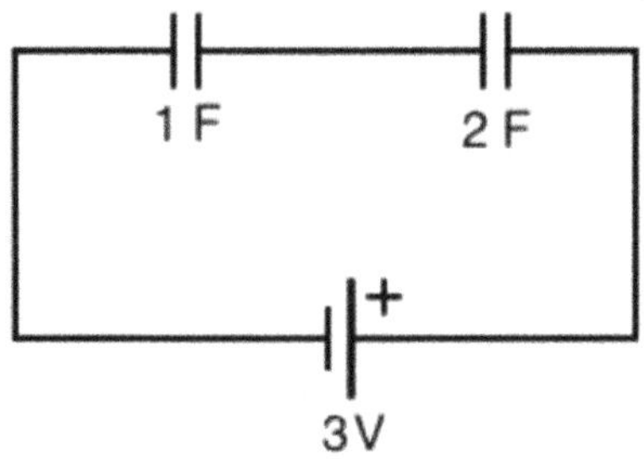

Figure M4.2.

5) Calculate the capacitance of a parallel plate capacitor with 2×2 square plates separated by a distance of 1 mm.

Reference

[1] Jones W 1980 Earnshaw's theorem and the stability of matter *Eur. J. Phys.* **1** 85–8

IOP Publishing

The Foundations of Electric Circuit Theory

N R Sree Harsha, Anupama Prakash and D P Kothari

Chapter 5

Electric currents

5.1 Special theory of relativity

As a matter of fact, we say there are some events that are absolute and some that are relative. This statement can be justified with a simple example. For person A, a spoon is at the left of a cup, but from the point of view of person B sitting opposite to person A, it is the other way around. Thus the position of any object is relative to the position of an observer. On the other hand, if the cup is filled to the brim with tea, all the observers will agree to this fact regardless of their position. Thus this is an absolute statement, independent of who makes the observation.

The concept of the absolute when applied to time and space facilitated Newtonian mechanics. According to Newton, absolute time and space, respectively, are independent aspects of objective reality. In 1687, Newton published his first book [1] on the laws of motion, which introduced the concept of 'absolute motion', 'absolute space' and 'absolute time'. He wrote, 'Absolute motion is the translation of a body from one absolute place into another; and relative motion, the translation from one relative place into another.' In particular, Newton thought of empty space as a fixed frame of reference through which objects experience motion.

In 1632, Galileo Galilei explicitly enunciated the principle of relativity [2]. His relativity hypothesis states that 'Any two observers at uniform motion with respect to one another will obtain the same results for all mechanical experiments.' Complying with Galilean relativity, Newton constructed his mechanics and said that for two inertial observers moving relative to each other the question 'which one is moving?' is unanswerable and meaningless, which he explained with the following example. An observer in uniform motion with respect to another cannot, without looking outside, determine whether he is at rest or not. And even if he looks outside, he cannot decide whether he or the other observer is in motion. However, the same question 'are we moving?' becomes meaningful once we specify a reference frame. This fact, stated in the 17th century, is relevant even today and is one of the postulates of Einstein's theories of relativity. Einstein extended the idea to include

doi:10.1088/978-0-7503-1266-0ch5

electrodynamics and called it 'the theory of relativity' in his famous 1905 paper 'On the electrodynamics of moving bodies' [3]. This theory changed our understanding of space and time.

Formally, Einstein's special theory of relativity is based on two postulates, which are:

(a) the principle of relativity.
(b) the light postulate.

The principle of relativity

The first principle, the principle of relativity, emphasizes that:

The laws of physics are the same in all inertial frames of references.

This means that the statement of a law of physics in one inertial frame of reference is exactly the same in any other inertial frame of reference. That is, when an observer on spaceship A and an observer on spaceship B conduct the same experiment (which is definite and gives repeatable results), they obtain the same result. According to Newtonian mechanics, all motion can be thought of as moving with respect to an empty space, which is considered to be at absolute rest. However, mechanical experiments indicate otherwise, which can be explained using a very simple example: a person travelling in a uniformly moving train on a straight track cannot tell from the movement of a toy car what the motion of the train is with respect to the ground. Thus no matter what the relative motion may be, as long as it is constant, the results are identical. Since the laws of mechanics are the same in all frames there is no preferred inertial frame of reference. Hence, there is no physically definable absolute rest frame. The same is true for electrodynamics. No experiment in electrodynamics carried out in a single frame will enable us to determine the velocity of our frame with respect to any other frame. In order to illustrate that the laws of physics assume the same form in all inertial frames of reference, let us consider Ohm's law, which says that the voltage across a resistor is proportional to the current flowing through it. Thus applying 3 V to a 1 ohm resistor results in a flow of 3 A of current through it. If this experiment succeeds in one inertial frame of reference, it will also give the same result in any other inertial frame of reference. Thus there is no indication that any inertial frame is intrinsically special and different from the others. Since absolute rest and absolute motion, if detected, would require a non-existent distinguished rest frame, one can conclude that no experiment that is conducted inside the frame can reveal the absolute motion of the observer.

The principle of relativity: the laws of physics are the same in all inertial frames of reference.

The three important conclusions are:

1. **Absolute motion cannot appear in any laws of physics.**
2. **All experiments run the same in all inertial frames of reference.**
3. **No experiment can reveal the absolute motion of the observer.**

The light postulate
The second principle, the light postulate, states that:

The speed of light in free space has the same value c in all inertial frames of references.

As measured in any inertial frame of reference, light is always propagated in empty space with a definite velocity c that is independent of the state of motion of the emitting body.

The second postulate is really a consequence of the first, because if Maxwell's equations hold in all inertial frames, then the only possible value for the speed of light is c. We shall see how this is true in chapter 8 when the full set of Maxwell's equations is formulated.

Maxwell, with his theory of electromagnetism [4], showed that electric and magnetic fields can give rise to a wave-like phenomenon and calculated its speed of propagation to be equal to the speed of light, thereby establishing that light is an electromagnetic oscillation. In the early part of the 19th century, the wave nature of light had been established by experiments such as interference, diffraction, etc. In order for light to propagate through empty space, like all other wave motions, it needed a medium. It was thought that empty space was filled with a mysterious medium called ether, which filled the whole of empty space and was not easily detected. The sole purpose of this was to act as a carrier for electromagnetic waves. He wrote another famous paper in 1864 in which the details of the luminiferous medium were less explicit and also indicated some uncertainty surrounding the precise nature of his molecular vortices [5]. He also recognized that light is an electromagnetic wave of very high frequency that propagates at a constant speed (c) of 299 792 458 m s^{-1}, which is independent of the state of motion of the emitting body in just one frame of reference, i.e. the rest frame of electromagnetic ether. The speed of light can be derived using Maxwell's four equations on electromagnetism, as discussed in the following chapters.

Einstein, however, with his theory of relativity, was sure that the laws of nature were the same for all inertial observers. He considered the constancy of the speed of light to be a law of nature. He then redefined the ideas of space and time in such a way as to fit in the constancy of the speed of light in all inertial frames of reference. One of the kinematic consequences of doing this is the so-called 'relativity of simultaneity'.

5.2 Relativity of simultaneity

The theory of the relativity of simultaneity states that:

Two events that are simultaneous in one inertial frame are not simultaneous in any other frame.

One of the most surprising features of special relativity is that a number of statements and results that we usually think of as absolute turn out to be observer-dependent. This

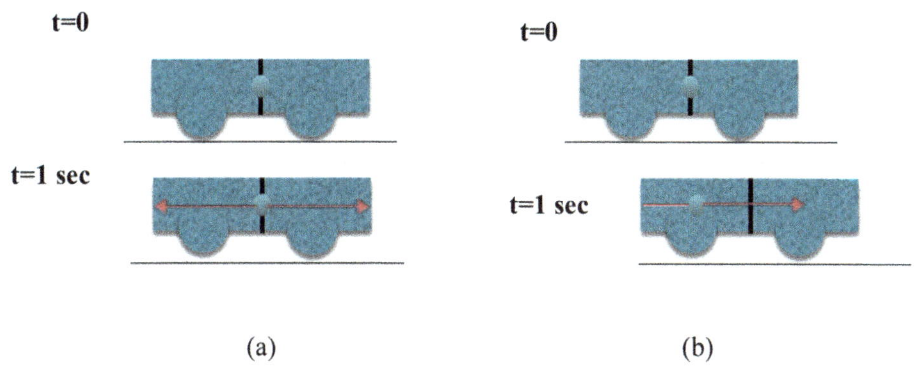

Figure 5.1. The train-and-embankment thought experiment: (*a*) from the reference frame of observer *A* and (*b*) from the reference frame of observer *B*.

concept can be explained with the help of a train-and-embankment experiment first suggested by Einstein in 1917. Let us consider a train travelling along a straight rail track with a constant velocity *v*, as indicated in figure 5.1(*a*). There are two observers, observer *A* standing exactly in the middle of the running train and observer *B* standing on a platform outside the train. A flash of light occurs at the center of the train exactly when the two observers *A* and *B* are standing in front of each other. According to observer *A*, who is stationary with respect to train, the front and rear part of the train are equidistant from the source of light and thus according to him the light will reach the front and back of the train at the same time, as shown in figure 5.1(*a*). However, according to observer *B*, the rear part of the train is moving towards the source of light and the front part of the train is moving away from it. Since the speed of light is constant regardless of the reference frame, the light moving towards the back of the train will have to travel less distance than the light moving towards the front portion of the train. Thus for observer *B* the flashes of light will strike the front and back ends of the train at different times, as shown in figure 5.1(*b*). This shows that events that are simultaneous with reference to the train are not simultaneous with respect to the platform.

Thus two events that are simultaneous in one frame have some time difference in another frame, and this difference increases with an increase in the relative speed between the two frames, and also increases with the distance between the two events along the direction of the relative motion.

Hence statements about space and time and the distances and duration of an event turn out to be relative and lead to the concepts of time dilation and length contraction.

5.3 Time dilation

This is based on the assumption that all observers measure the same value for the speed of light. This means that time runs at different rates for different inertial frames. There is no absolute time; time only has a relative meaning, as it depends on the relative velocity between two objects. One can observe this strange effect only when velocity is near the speed of light *c*.

This can be explained with the help of Einstein's light clock. The famous clock consists of two mirrors placed facing each other in opposite directions with a flash of light bouncing between them. Suppose that the distance between these two mirrors is *d* km and that the time taken by light to travel this distance is 1 s, as shown in figure 5.2(*a*). If a person *A*, who is at rest with respect to the clock, is holding it, with the light bouncing between two mirrors and covering a distance *d* km in 1 s, then the speed of light *c* is given by equation (5.1):

$$c = \frac{d}{t} = \frac{d}{1} \text{ km s}^{-1}. \quad (5.1)$$

As per the principle of relativity, if an inertial observer *X* measures the speed of a flash of light to be *c* then another inertial observer *Y*, who chases the flash of light at a very high speed of, say, 80% of *c* should also only measure the speed of light to be *c*. Speed is just a measure of distance covered in a given time. In order to agree on the speed of light, different observers might have to disagree about distance and time.

Let us consider an inertial frame *S*, sufficiently far away from other external interactions. Now suppose that a person *A* is travelling in a spaceship with the same light clock with a velocity of *V* km. The clock is travelling at the same speed as the spaceship and in the same direction.

Scenario 1

From the point of view of person *A*, who is moving along with the clock w.r.t. frame *S*:

$$\text{time } t_1 = 1 \text{ s, speed of light } c = d/t_1 = d/1 \text{ (km s}^{-1}\text{)}.$$

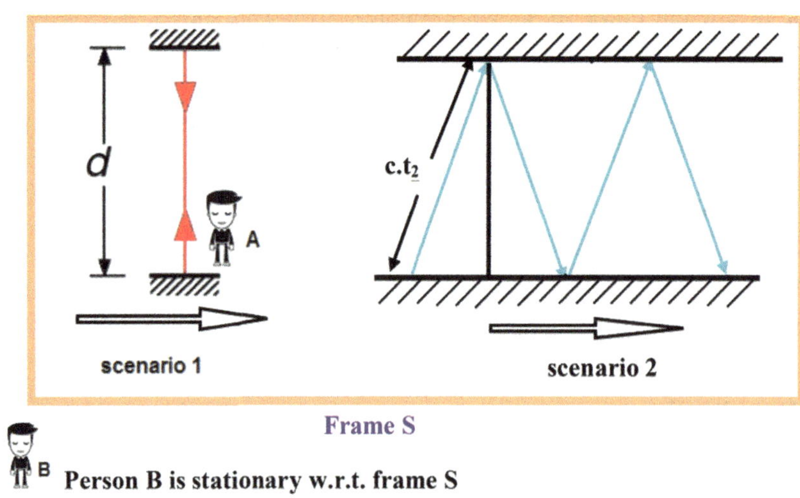

Figure 5.2. Einstein's clock: (*a*) w.r.t. person *A* and (*b*) w.r.t. person *B*.

Scenario 2
From the point of view person B who is stationary w.r.t. frame S:

$$\text{speed of light } c = D/t_2'.$$

From the point of view of person A the light is travelling up and down, but from the point of view of person B, who is stationary with respect to frame S, it will seem that the light follows a v-shaped path, as shown in figure 5.2(*b*). The v-shaped path is longer than the straight up and down path. Thus for person B the light has to cover a distance $D = c \cdot t_2$ ($>d$) km with the same speed.

From Maxwell's equations, the speed of light is not affected by the speed of the object emitting it, in other words, it is always constant. Hence, from figure 5.2(*b*), on applying Pythagoras' theorem, it is evident that

$$(c \cdot t_2)^2 = d^2 + (v \cdot t_2)^2$$

$$c^2. t_2^2 - v^2 \cdot t_2^2 = d^2$$

$$t_2 = \frac{d}{\sqrt{c^2 - v^2}} = \frac{d}{c\sqrt{1 - v^2/c^2}}.$$

Since $d/c = t_1$,

$$t_2 = \frac{t_1}{\sqrt{1 - \frac{v^2}{c^2}}}.$$

Thus, person B feels that person A's clock is running more slowly than his clock. Hence, moving clocks run more slowly than similar clocks at rest.

The closer the spaceship approaches the speed of light, the slower the time inside the ship will flow for observer B. If the speed of the ship is exactly equal to the speed of light, then for observer B time inside the ship will stop altogether. In fact, person A, who is moving along with the clock, will not notice this and time flows normally for him. This is called time dilation. Thus to understand the concept of relativity we should stop thinking that time is constant for everyone; time is relative.

There is no universal clock in the Universe ticking off the same universal time for everyone.

5.4 Rods moving perpendicularly to each other

Rods that are placed perpendicularly to their motion do not change their perpendicular length. In order to understand this phenomenon, consider a hypothetical experiment involving two moving rods placed perpendicularly to each other. Construct two hypothetical rods L_1 and L_2 and verify that that their lengths match. Now put black paint on both the ends of rod L_1 and separate the rods to a certain distance, as shown in figure 5.3.

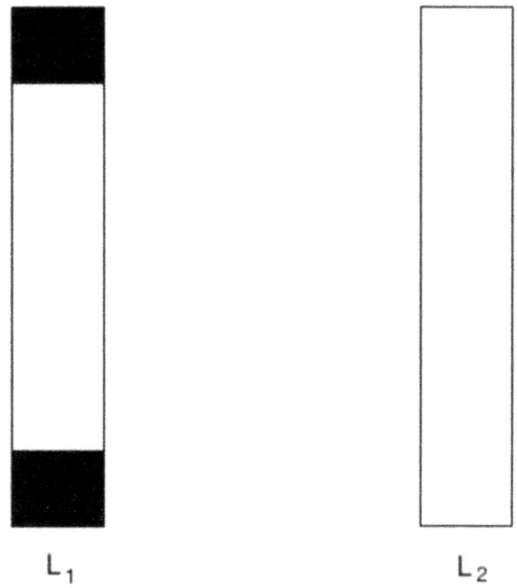

Figure 5.3. Rod L_1, which has the same length as rod L_2, is painted black at its ends.

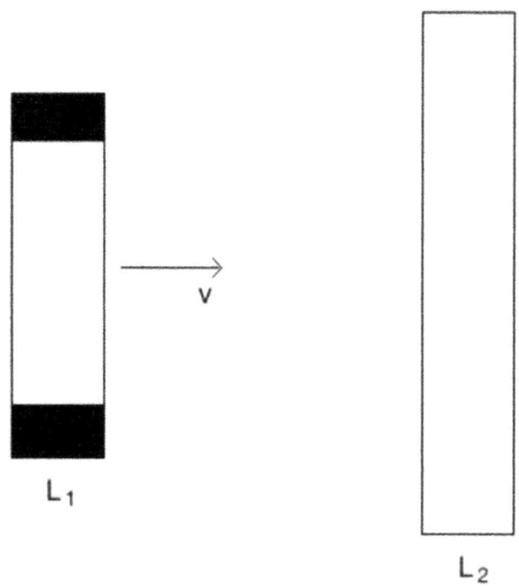

Figure 5.4. The situation as viewed from a frame attached to rod L_2.

Next, suppose that we move rod L_1 with a uniform velocity v towards rod L_2, as shown in figure 5.4. Let us analyse this movement from different frames of reference.

Frame attached to rod L_2

For the frame attached to rod L_2, the rod L_1 is moving towards the rod with a velocity v, as shown in figure 5.4. Let us suppose for the sake of the argument that rods placed perpendicularly to their motion shrink by some factor. Hence, we observe from the frame attached to rod L_2 and conclude that the length of L_1 is shrunk and therefore the paint will be deposited on rod L_2.

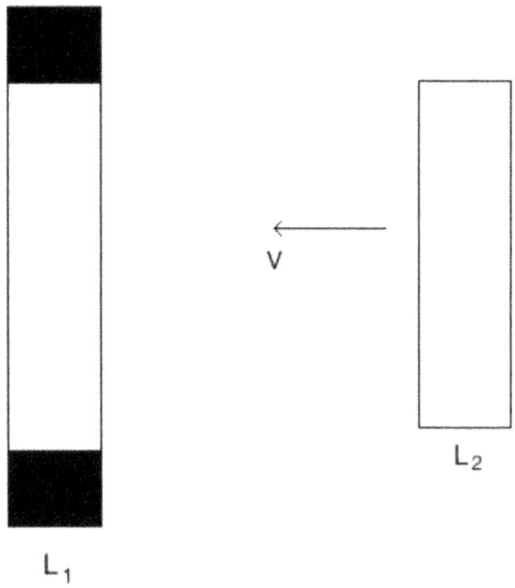

Figure 5.5. The situation as viewed from a frame attached to rod L_1.

Frame attached to rod L_1

Let us now analyse the same situation with respect to a frame attached to the rod L_1. With respect to this frame, rod L_2 is moving towards rod L_1 with a velocity v, as shown in figure 5.5. Hence, rod L_2 should shrink and the paint should not be deposited on rod L_2.

Thus, we are led to a contradiction. The paint should either be deposited on the rod or it should not. By the principle of relativity, both of the frames are equally good for analysing the situation. The only way to resolve this contradiction is to assume that the rods placed perpendicularly to their velocity do not shrink in their perpendicular length.

5.5 Length contraction

The length of the objects parallel to their motion contract as viewed from a rest frame. Let us apply the same concept to Einstein's clock by rotating it by 90° in the clockwise direction, as shown in figure 5.6.

From the point of view of person A the clock is stationary and the time taken by the light to travel from mirror X to mirror Y and back to mirror X is given by

$$t_0 = \frac{2d}{c}. \tag{5.2}$$

From the point of view of person B the clock is moving at a speed of v km s^{-1} and also each tick of the clock is dilated and the time between each tick is

$$t_1 = \frac{t_0}{\sqrt{1 - \frac{v^2}{c^2}}}. \tag{5.3}$$

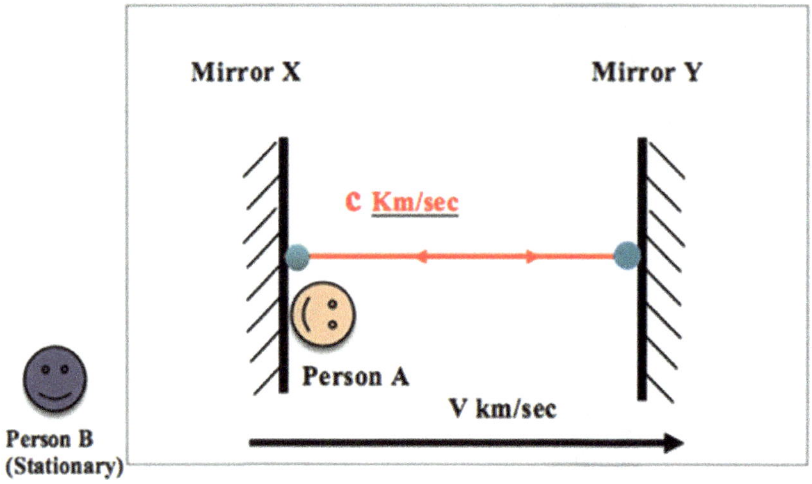

Figure 5.6. Einstein's clock in the horizontal state.

Let us assume that the length of the distance between mirrors changes to d_1. Then the total time the light particle takes to go from X to Y is

$$T_1 = \frac{d_1}{c - v}. \tag{5.4}$$

The total time the light particle takes to travel from Y to X is

$$T_2 = \frac{d_1}{c + v}. \tag{5.5}$$

The total time of the tick of the clock is then given by $t_1 = T_1 + T_2$. On putting in the value of T_1 and T_2 from equations (5.4) and (5.5), we get that

$$t_1 = \frac{2d_1c}{c^2 - v^2}. \tag{5.6}$$

We have already seen that

$$t_1 = \frac{t_0}{\sqrt{1 - \frac{v^2}{c^2}}}.$$

Hence, we have

$$t_1 = \frac{2d_1c}{c^2 - v^2} = \frac{t_0}{\sqrt{1 - \frac{v^2}{c^2}}} = \frac{2d}{c\left(\sqrt{1 - \frac{v^2}{c^2}}\right)}. \tag{5.7}$$

From equation (5.7), the length of the distance between the mirrors from person B's point of view is given by

$$d_1 = d\left(\sqrt{1 - \frac{v^2}{c^2}}\right). \tag{5.8}$$

This shows that $d_1 < d$.

Thus there has been length contraction along with time dilation for the speed of light to be constant. This means that the higher the velocity, the smaller the length. However, it should be noted that only lengths parallel to the relative velocity are contracted. The lengths perpendicular to it remain the same.

This same principle of length contraction is applied to an empty box of volume V, as shown in figure 5.7(*a*). An observer passing this box at a speed of s finds a reduction in the size of the box by a factor in the direction of motion due to Lorentz contraction, as shown in figure 5.7(*b*).

Now consider the same box to be filled with n particles as shown in figure 5.8(*a*). In this case, if an observer passes by with a speed of s, he finds the box to be Lorentz contracted, but there is no change in the number of particles, as shown in figure 5.8(*b*).

Since the number of particles inside is unchanged, it can be concluded that the volume density seen by this observer is larger than the volume density seen by an observer at rest with the box. This concept, when applied to a current-carrying conductor, modifies the existing equation of electric current.

We shall now see how forces transform when we go from one inertial frame to another. We know that the transverse momentum should be the same in both inertial frames S and S'. Suppose that S' is moving in the positive x-direction with respect to

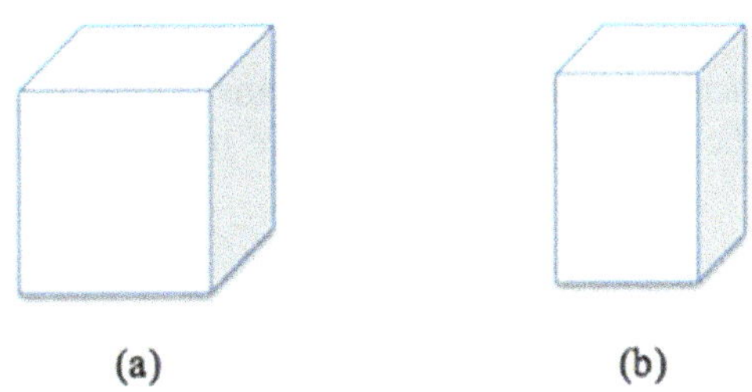

Figure 5.7. (*a*) An empty box. (*b*) To an observer passing by, the same box appears to be Lorentz contracted.

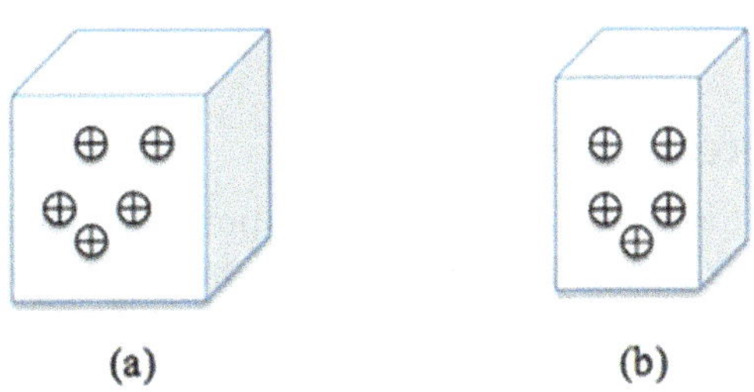

Figure 5.8. (*a*) A box containing a certain number of particles. (*b*) To an observer moving by, the box is Lorentz contracted, and thus the number density seen by this observer is greater than that seen by an observer at rest with the box.

S at a speed v. If p_x denotes the transverse momentum on a particle in the S frame, we have that

$$\Delta p_x = F\Delta t.$$

In the S' frame, we should have that

$$\Delta p'_x = F'\Delta t'.$$

For the corresponding time intervals, we must have that $\Delta p_x = \Delta p'_x$. Since our particle is initially at rest with respect to S', we must expect that

$$\Delta t = \frac{\Delta t'}{\sqrt{1 - \frac{v^2}{c^2}}}.$$

And since $\Delta p_x = \Delta p'_x$, we must have that

$$F' = \frac{F}{\sqrt{1 - \frac{v^2}{c^2}}}.$$

Let us now consider the relativistic addition of velocities u and v. Suppose that in a frame S' a particle is moving with a velocity v. Next, suppose that frame S' is itself moving with a velocity u in the same direction with respect to another frame S. What is the velocity of the particle in the frame S? Newtonian mechanics naively says that the velocity w of the particle in the frame S is given by

$$W = u + v.$$

This result is relativistically incorrect, as this means velocities greater than the speed of light c. However, the correct equation that we must use is

$$W = \frac{u + v}{1 + \frac{uv}{c^2}}.$$

If we use the above equation to find out the net velocity of the particle with respect to frame S, we can see that it cannot exceed the speed of light. These results will be useful when we derive the equation for the magnetic field created around a current-carrying conductor.

5.6 Modified expression of current

We know that current is flow of charge per unit time. In 1900 Paul Drude came up with a model [6], which explains the phenomenon of conduction in metals. The Drude model considers the metal to be formed of a mass of positively charged ions from which a number of 'free electrons' were detached. These free electrons have some average thermal energy ($mv_T^2/2$), but they pursue random motions through the metal so that $v_T = 0$ even though $v_T^2 \neq 0$. Under the application of a field, electrons experience a force and are accelerated towards the end at higher potential. Due to this acceleration, the velocity of the electrons increases for only a short interval of time as each accelerated electron suffers frequent collisions with positive ions and loses its kinetic energy. After each collision, an electron starts afresh in a random direction, and is again accelerated and again loses its gained kinetic energy in another collision. This extra velocity gained by the electron is lost in subsequent collisions and the process continues until the electron reaches the positive end of the conductor. Under the effect of electric field inside the conductor, free electrons acquire a drift velocity, which is superimposed on the thermal motion and drift towards the positive end of the conductor. If τ is the average time between two successive collisions and E is the strength of the applied electric field, then the force on the electron due to applied electric field is

$$F = eE, \tag{5.9}$$

where e is the amount of charge on the electron, and if m is the mass of electron, then the acceleration produced is given by equation (5.10):

$$a = \frac{eE}{m} \qquad (\because F = ma). \tag{5.10}$$

Since the electron is accelerated for an average time interval (τ), an additional velocity (v_d) acquired by the electron is given by equation (5.11)

$$v_d = a\tau. \tag{5.11}$$

Putting the value of a into equation (5.11) from equation (5.10):

$$v_d = \frac{eE}{m}\tau. \tag{5.12}$$

This small velocity, which is imposed on the random motion of electrons in a conductor on the application of electric field, is known as *drift velocity* (equation (5.12)).

Let us consider a current-carrying conductor of length L and cross-section area A. If there is n number of free electrons per unit volume of the conductor when there is no current flowing through it, then the total number of free electrons in the conductor will be nAL. If each free electron has a charge of e then the total charge of the free electrons is $enAL$. If the length of the wire L is the distance travelled by free electrons in 1 s, then the current is given by equation (5.13):

$$i = \frac{neAL}{1\text{s}}(\text{amp}) = i = neAv_d, \tag{5.13}$$

where $v_d = \dfrac{L}{1\text{ s}}$ is the drift velocity of free electrons.

Since $i \propto v_d$, the drift velocity of the electrons approaches infinity when the current through the conductor approaches infinity. This statement is inconsistent with the second postulate of the special theory of relativity, which states that c is the maximum speed at which all matter and information in the Universe can travel.

In order to overcome this issue, let us apply the Lorentz contraction theory to a current-carrying conductor. As discussed above, we know that if charges are in motion, their volume density changes as the time component of a four-vector because of Lorentz contraction. However, *Lorentz contraction exists not only when there is relative motion between the charge and the conductor, but also when the current flows through the conductor.*

Let us consider that n represents the number of electrons per unit volume in a conductor when there is current flowing through it. Since n changes as a time component of a four-vector due to Lorentz contraction, it can be written as shown in equation (5.14)

$$n = n^{\dagger}\gamma, \tag{5.14}$$

where $n^{\dagger}$ represents the number of free electrons per unit volume when the conductor is at rest and when it does not carry any current, and $\gamma = \dfrac{1}{\sqrt{1 - \dfrac{v_d^2}{c^2}}}$.

On placing the value of n from equation (5.14) into equation (5.13), we obtain the modified expression for current as given in equation (5.15)

$$i = \frac{n^{\dagger}eAv_d}{\sqrt{1 - \dfrac{v_d^2}{c^2}}} \tag{5.15}$$

$$i = \lambda^{\dagger}v_d,$$

where $\lambda^{\dagger} = \dfrac{n^{\dagger}eA}{\sqrt{1 - \dfrac{v_d^2}{c^2}}} = neA,$

where $\lambda^{\dagger}$ is the charge per unit length of the conductor, which is a constant for a given conductor.

However, the modified equation of current indicates that if the drift velocity of electrons approaches the speed of light, the current in the conductor approaches infinity, which is in accordance with the consequences of the special theory of relativity. Once the concept of current in a conductor is clearly understood, it is important to analyze the relation between current and voltage in electrical circuits.

5.7 Ohm's law

The German physicist Georg Ohm [7], who was inspired by Fourier's work on heat conduction, conducted an experiment on wires of varying length, diameter and material, along with a thermocouple (voltage source) and a galvanometer (to measure current), and stated that the electromotive force at a given point of a conductor was proportional to the gradient of the electroscopic force or tension (identified as charge density). Another German physicist named Gustav Kirchhoff [8] realised that the equation given by Georg Ohm correctly gave the distribution of currents in a conductor but was incompatible with electrostatics. He then reformulated the equation given by Georg Ohm into another equation, as shown in equation (5.16), and stated that the current density is directly proportional to the gradient of the electrostatic potential.

$$\overline{J} = \frac{\overline{E}}{\rho}, \tag{5.16}$$

where $\overline{E}$ is the electric field vector with units of volts per meter, $\overline{J}$ is the current density vector with units of amperes per unit area, and ρ is the resistivity with units of ohm-meters. The above equation is sometimes written as $J = \sigma E$ where σ is the conductivity, which is the reciprocal of ρ.

Kirchhoff (as given by equation (5.17)) also formulated the scalar form of the vector equation (4.16) and he named these two equations Ohm's law, after Georg Ohm

$$I = \frac{V}{R}, \tag{5.17}$$

where I is the current through the conductor in units of amperes (analogous to J of Ohm's law, which has units of amperes per unit area), V is the potential difference measured across the conductor in units of volts (analogous to E of Ohm's law, which has units of volts per meter) and R is the resistance of the conductor in units of ohms (analogous to ρ of Ohm's law, which has units of ohm-meters). Thus Ohm's law states that the current through a conductor between two points is directly proportional to the potential difference across the two points. The constant of proportionality, the resistance, is constant, and is independent of the current.

Gustav Kirchhoff generalized the work of Georg Ohm and presented *Kirchhoff's laws*. These Kirchhoff laws are very widely used to solve various electric circuits. The basic electric circuit is a closed loop with at least one source of voltage (thus providing an increase of potential energy) and at least one element (a place where potential energy decreases). An increase of potential energy in a circuit causes a charge to move from a lower to a higher potential and a decrease of potential energy can occur due to heat lost in electrical resistance. Then, according to equation (5.17), when a voltage of 1 V is applied to a resistor of 1 Ω it will cause a current of 1 A in a certain direction, as shown in figure 5.9. Ohm's law not only gives the magnitude of the current flowing through the conductor, but it also indicates its direction.

In addition to voltage and current, there is another measure of free electron activity in a circuit: power. Electric power (P) in a circuit is the amount of energy that is absorbed or produced per unit time within the circuit. A source of energy such as a voltage will deliver power while the connected load such as a resistor converts it into heat. If a current I flows through a resistor R, the amount of power P that is dissipated into heat is given by the well known formula $P = I^2R$ (watts). This well known formula tells us about the magnitude of the power dissipated but does not give any information about the flow of electrical energy. However, it is often misunderstood that electrical energy flows in the same direction as the direction of

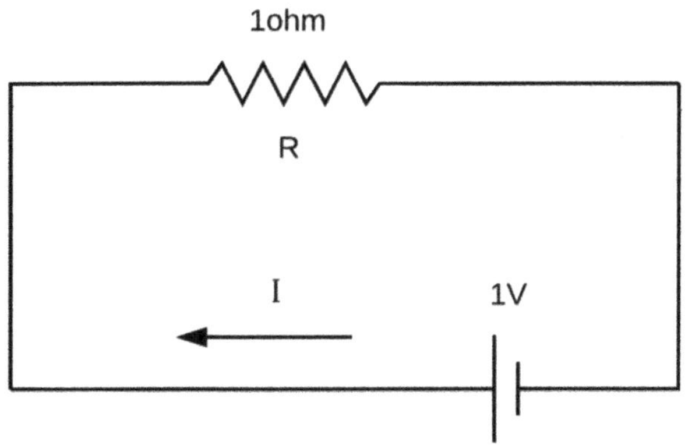

Figure 5.9. A simple electric circuit.

the current. In order to address this issue it is important to understand the Poynting theorem, which is discussed in the next section.

5.8 Application of the Poynting vector to a simple DC circuit

In 1884 the British physicist John Henry Poynting developed the Poynting vector, which describes the direction and magnitude of the flow of electromagnetic energy flux density and is an integral part of the Poynting theorem [9, 10]. The Poynting vector $\vec{S}$ is given by

$$\vec{S} = \frac{1}{\mu_0}\left[\vec{E} \times \vec{B}\right].$$

We shall derive the theorem later in chapter 8, since it requires a full understanding of Maxwell's equations. In this section, we shall apply the Poynting vector to a DC electric circuit and see how energy is transferred from a voltage source to a resistor.

In order to apply the Poynting vector to a DC circuit, we first need to understand the electric fields in the vicinity of the circuit. Sommerfeld discussed the origin of the electric field inside a conductor in a DC circuit and discovered that this field is created by the charges present on the surface of the conductor, which he called *surface charges*. There are two different types of surface charges.

- Type 1 surface charges occur at the boundary of conductors with different resistivities.
- Type 2 surface charges occur at the boundary of the conductor. They are present on the surface between the conductor and the air.

5.8.1 Type 1 surface charges

Consider two conductors with resistivities ρ_1 and ρ_2 with the same cross-sectional area A and connected in series, as shown in figure 5.10.

We know that with the same current density **J** must flow through this series combination when connected across a potential difference. Let the electric field in the first material be $\mathbf{E}_1$ and let the electric field in the second material be $\mathbf{E}_2$. For the

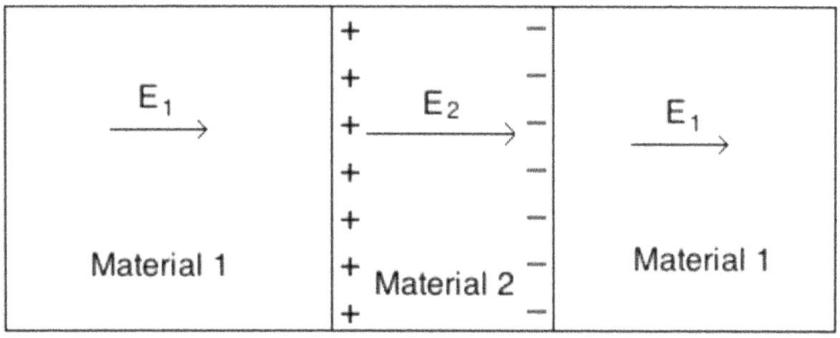

Figure 5.10. Materials with different resistivities connected in series.

same current density to flow in the two different materials, we must have from Ohm's law $\vec{J} = \vec{E}/\rho$ that

$$\frac{\vec{E}_1}{\rho_1} = \frac{\vec{E}_2}{\rho_2}. \tag{5.18}$$

Gauss's law relates the surface charge density σ produced at the interface to the electric fields in both of the materials. The tangential component of the electric field remains the same, but the normal component of the electric field experiences a change given by

$$\vec{E}_2 - \vec{E}_1 = \frac{\sigma}{\varepsilon_0}. \tag{5.19}$$

Let us assume that the current is I. From equations (5.18) and (5.19) we get that

$$\sigma = \frac{1}{A}\varepsilon_0\left(\rho_2 - \rho_1\right). \tag{5.20}$$

Applying equation (5.20) to the left and the right boundaries, we see that

$$\sigma_{\text{left}} = \frac{1}{A}\varepsilon_0\left(\rho_2 - \rho_1\right)$$

$$\sigma_{\text{right}} = \frac{1}{A}\varepsilon_0\left(\rho_2 - \rho_1\right)$$

Hence, we conclude that the positive charges are created on the left boundary and the negative charges are created on the right boundary, as shown schematically in figure 5.10. These surface charges increase the electric field in materials with high resistivity to maintain the same current density.

5.8.2 Type 2 surface charges

These surface charges are present on the surface of the conductor and provide an electric field along the direction of the current density. A typical distribution of surface charges on a resistor is shown schematically in figure 5.11.

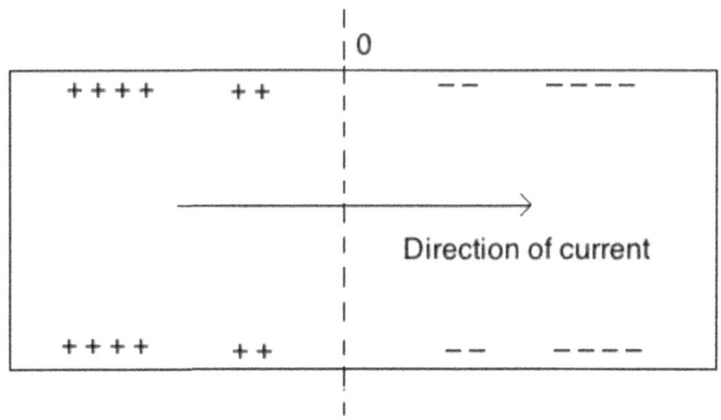

Figure 5.11. A typical charge distribution on the surface of a resistor.

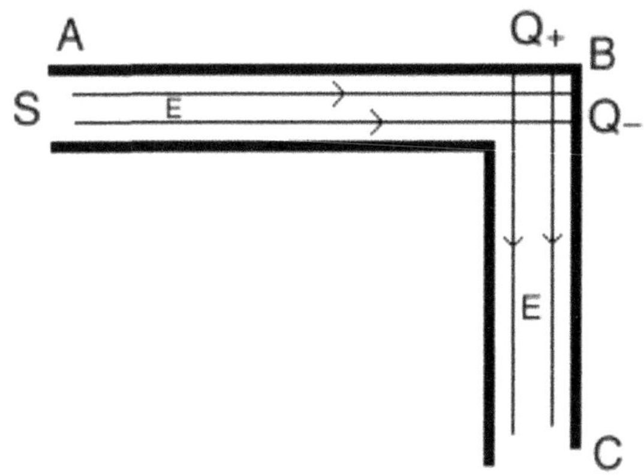

Figure 5.12. A portion of a circuit shown to describe the magnitude of charges required to bend the electric field as shown.

These charges provide the electric field necessary for the electrons to bend along the path of the conductor and also confine the currents to the conductor without allowing them to escape. The accumulation of these charges takes place very quickly in the transient part of closing a switch in a DC circuit and the process is completed as soon as the total electric field points along the wire at every point inside the circuit. The distribution of this type of surface charges is very complex because it involves a detailed description of the geometry of the circuit. Let us now try to understand the magnitude of these surface charges required to make an electric current I turn along a corner of a circuit. Consider a portion of a circuit made of a material with conductivity σ and a cross-sectional area S, as shown in figure 5.12.

Inside the conductor and away from the bend, the electric field is parallel to the wire. The total flux of the electric field $\mathbf{E}$ in the section AB of the wire is ES. These electric field lines are terminated at a point B in the circuit. From Gauss's law, the negative charge Q_- capable of terminating the flux ES is given by

$$Q_- = -\varepsilon_0 ES.$$

Since $\vec{J} = \sigma\vec{E}$ and $J = I/A$, we have

$$Q_- = -\frac{\varepsilon_0 I}{\sigma}.$$

Similarly, the positive charge that will start the flux ES of the electric field along the direction BC of the portion of the circuit is

$$Q_+ = \frac{\varepsilon_0 I}{\sigma}.$$

For a copper wire with $\sigma = 5.7 \times 10^7$ mho m^{-1}, we have that

$$|Q_+| = |Q_-| = 1.5 \times 10^{-19}\ IC.$$

For a current of 100 A, we have that $|Q_+| = |Q_-| = 1.5 \times 10^{-17}C$. This is approximately the total charge on 100 electrons. Hence, we see that extremely small macroscopic charge distributions are sufficient to change the direction of electric field inside a current-carrying conductor. This derivation of the estimate of the order of the surface charges around a corner of a portion of a circuit was first given by W G V Rosser in a brilliant note [11].

We shall, however, take a particularly simple circuit to demonstrate how these surface charges transfer energy from the voltage source to the resistor.

The Poynting vector helps in conceptualizing and thus not only quantifies but also gives the direction of the flow of the electromagnetic energy in circuits. The question of the direction of the electrical energy can alternatively be asked like this—in what way is the electric potential energy of electrons in the battery converted into heat energy in the resistor connected across it [12]?

In order to understand this, it is important to first understand the behavior of an electric field in the vicinity of the electric circuits shown in figure 5.13.

The simple closed circuit has an ideal battery connected with a homogeneous wire with resistance r. The steady current in the resistive wire clearly implies an electric field. The gradient in the density of the surface charge provides the axial field E_t within the conductor, which guides the movement of the conduction electrons.

As shown in figure 5.11, the vector $\overline{E}$ has two components:

- $\overline{E}_n$ perpendicular and outward (or inward) to the wire
- $\overline{E}_t$ along the wire in the direction of current

The normal component $\overline{E}_n$ changes along the circuit; in magnitude, reflecting the gradient of the surface charge, and in direction, reflecting the presence of surface charges. The magnitude of the tangential component $\overline{E}_t$ remains the same. These electric fields are created by the surface charges maintained by the battery as well as the charges present on the plates of the battery. These surface charges are created in the transient state while making an electric connection.

Now consider a circuit consisting of an ideal DC source connected across a resistor R with ideal homogeneous wires. Since the resistance of the wire is zero, Ohm's law requires that there is no gradient of the surface charges on the wire, resulting in $\overline{E}_t = 0$. The resultant electric field around the ideal wire is therefore

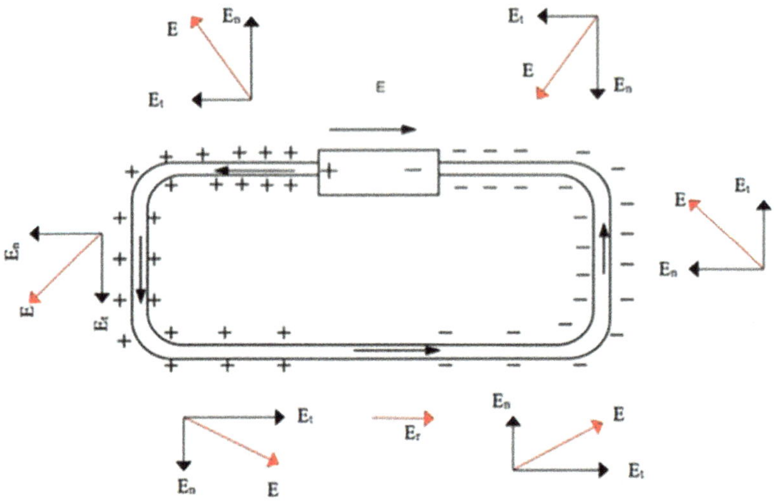

Figure 5.13. The electric field $\overline{E}$ along and outside the closed DC circuit. The case of ideal wires, with the resistor having uniform resistance R, and an ideal source with no internal resistance is considered.

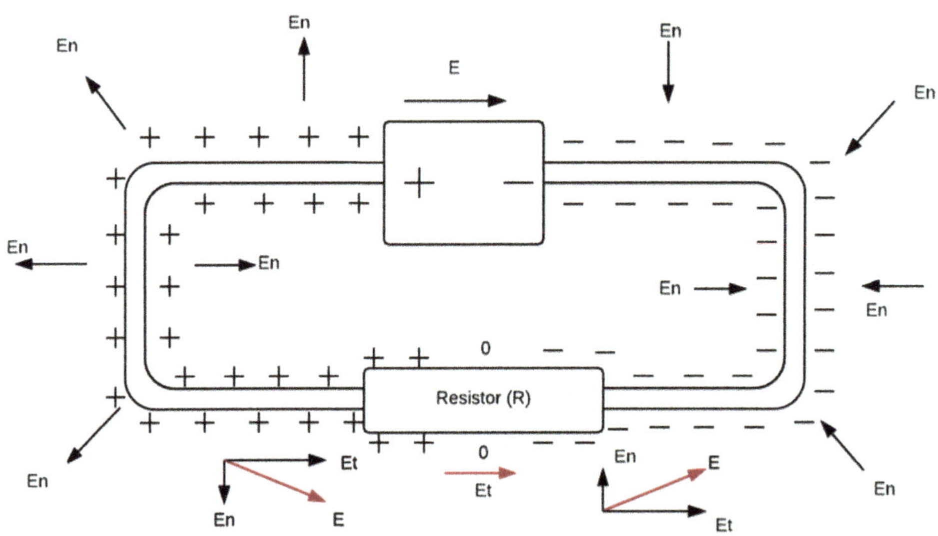

Figure 5.14. The rotation of the electric field $\overline{\mathbf{E}}$ (E_n, E_t) outside and along a closed DC circuit. Inside the wire the electrical field is collinear with the wire (E_t). The case of a homogeneous wire with resistance r and an ideal source with no internal resistance is considered.

perpendicular to the wire and directed outside it ($\overline{E} = \overline{E}_n$). On adding a resistor to such a circuit, the surface charge will cause the electric field $\overline{E}_n$ perpendicular to the resistor and gradient of the charges, which is present in the resistor results in the axial electric field $\overline{E}_t$ in the resistor. The resultant field $\overline{E}(= \overline{E}_n + \overline{E}_t)$ gradually turns along the resistor, as shown in figure 5.14

In summary, $\overline{E}$ is directed from the positive to the negative terminal within the battery. It is perpendicular to and outward from the surface for the ideal wires connecting the battery to the resistor, and continuously flips and changes in magnitude outside of the resistor. Thus the $\overline{E}$ vector reverses by 180° along the outer part of the circuit. This knowledge about $\overline{E}$ and the current direction enables us to determine the energy flux in the closed circuit and thereby determine the direction of the energy flow in the circuit.

The direction of the flow of electrical energy in the circuit is given in figure 5.15 and can be explained with the help of the Poynting vector. In order to apply the Poynting vector, it is important to estimate the direction of the magnetic and electric fields in the circuit. The direction of the magnetic field around a current-carrying conductor is determined by the direction of the current and the nature of the electric field is determined as per the above discussion.

Figure 5.15 shows $\overline{E}$ and $\overline{B}$ at different points along the circuit. The vector product shows the direction of the Poynting vector $\overline{S}$. In the battery, the Poynting vector is outward, indicating the direction of the energy flow outside the battery. This is due to the direction of the current, which is opposite to that of the electric field (caused by the charges present on the plates of the battery) created in the battery. In the vicinity of the conducting wires and next to the positive terminal of the battery, $\overline{S}$ is parallel to the wire. However, $\overline{S}$ is directed *from* the battery on both sides of the battery.

In the case of the resistor R, the change of direction of $\overline{E}$ outside the resistor causes $\overline{S}$ to change as well, with it gradually changing from being parallel to the resistor to be perpendicular to it.

Figure 5.15 demonstrates the rotation of the Poynting vector, which represents the energy flux from the generator along the wires, with this eventually arriving at the resistor and delivering energy to it. The result is that the electromagnetic energy flux is

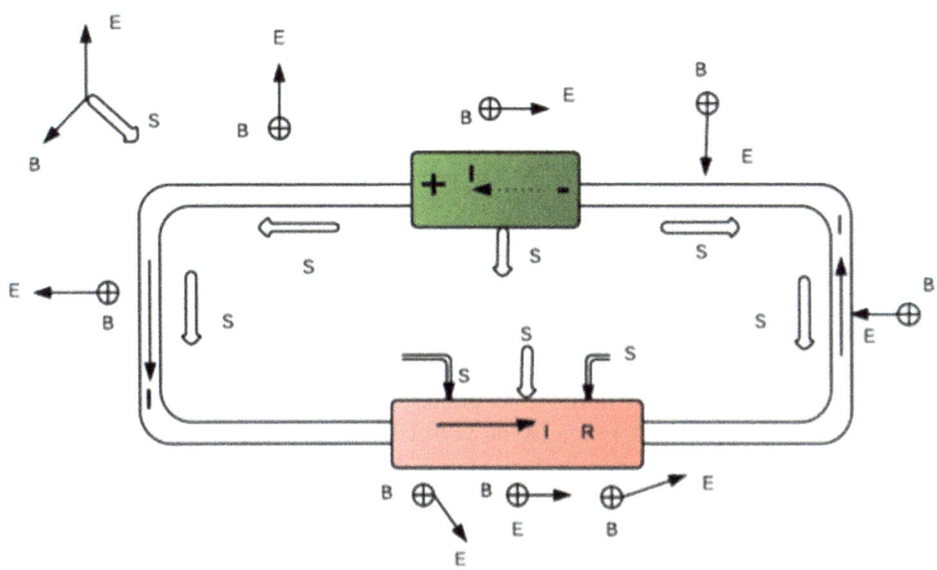

Figure 5.15. The electric and magnetic fields, $\overline{E}$ and $\overline{B}$, and the Poynting vector $\overline{S}$ along the closed DC circuit. The case of ideal wires, a homogeneous resistor R and an ideal source with no internal resistance is considered.

always directed from the source to the resistor and it never returns. Electrical energy is emitted into space in all directions and the resistor receives the energy from the space surrounding it. The electrical energy, surprisingly, is not carried inside the ideal wires because there is no electric field parallel to the wire in the case of an ideal wire.

The Poynting vector concept gives the direction of the energy flow and the Poynting theorem helps in estimating the expression of the energy flow rate. For an infinitely long wire carrying current I, the magnitude of the magnetic field B on the surface of a cylindrical resistor R of radius r_0 and length L is given by equation (5.21):

$$B = \frac{\mu_0 I}{2\pi r_0}. \tag{5.21}$$

The electric field E across this cylindrical resistor is given by equation (5.22):

$$E = \frac{IR}{L}. \tag{5.22}$$

The flux of the Poynting vector through the surface A of the resistor is given by equation (5.23):

$$\overline{S} = \frac{1}{\mu_0} \oint_A \left[\overline{E} \times \overline{B}\right] \mathrm{d}A \tag{5.23}$$

$$S = \frac{1}{\mu_0} EBA = I^2 R. \tag{5.24}$$

Equation (5.24) represents the usual form of the power loss in a resistor with a resistance R and a current I flowing through it.

Special case: short-circuiting a constant voltage source

As a special case let us consider the load resistance $R = 0$ (short circuited). In the previous case (shown in figure 5.14), the perpendicular component of the electric field was in one direction at the negative terminal of the battery, while it was in the opposite direction at the positive terminal, and the perpendicular component of the electric field outside the resistor changed its direction by 180° from one end of it to the other. Thus, the resistor acted as a device to change the direction of the perpendicular component of the electric field created by the battery. If the resistor in figure 5.14 were replaced by a short circuit, the direction of the electric field could not be changed, since there could not be any tangential component in an ideal conductor. Hence, the voltage source would try to send a very large amount of current through the ideal conductor to create a tangential component of electric field. This would result in damage of the voltage source. Hence, a constant voltage source should not be short-circuited.

Conclusion

In summary, we have seen that special relativity shares a very intimate relationship with electromagnetism. One cannot fully understand the behavior of electric circuits without understanding special relativity. We have also presented some proofs of already-established facts in a slightly different way so that the reader can have a strong understanding of the concepts.

Exercises

MCQs

1) If a constant force F acts on a particle, its acceleration will:
   a. Remain constant
   b. Gradually decrease
   c. Gradually increase
   d. None of the above
2) Which of the following quantities related to a particle has a finite upper limit?
   a. Kinetic energy
   b. Momentum
   c. Speed
   d. None of the above
3) In an electric circuit containing a battery, the negative charge inside the battery:
   a. Always goes from the positive terminal to the negative terminal
   b. May go from the negative terminal to the positive terminal
   c. Always goes from the negative terminal to the positive terminal
   d. None of the above
4) A resistor R is connected to an ideal battery. As the resistance of the resistor R increases, the power dissipated in the resistor:
   a. Increases
   b. Decreases
   c. Remains constant
   d. None of the above
5) Consider an electric circuit with a voltage source V connected to a resistor R with ideal conducting wires (zero resistance). Which of the following is true?
   a. Most of the energy transfer from the voltage source to the resistor occurs through the ideal conducting wires
   b. Most of the energy is transported to the resistor parallel to and just outside the conducting wires
   c. Energy is not transferred from the voltage source to the resistor
   d. The current given by the voltage source is consumed in the resistor with the help of the conducting wires

Problems

1) Calculate the magnitude of the drift velocity of the electrons when 5 A of current exists in a material with a 2 mm^2 cross-section. The number of free electrons in 1 cm^2 of that material is 8.5×10^{22}.

2) Calculate the resistance of an aluminum wire of length 100 cm and cross-sectional area 2.0 mm^2. The resistivity of aluminum is 2.6×10^{-8} Ω m.
3) Find the equivalent resistance between points A and B in figure M5.1.

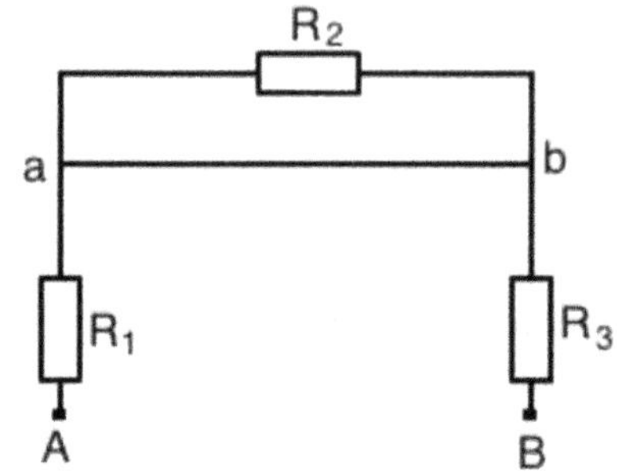

Figure M5.1.

4) Find the equivalent resistance between points A and B in figure M5.2.

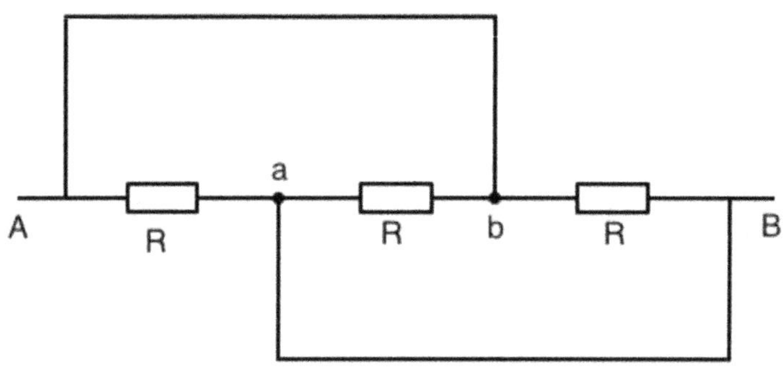

Figure M5.2.

5) Find the charge on one of the plates of the parallel plate capacitor C in the circuit shown in figure M5.3.

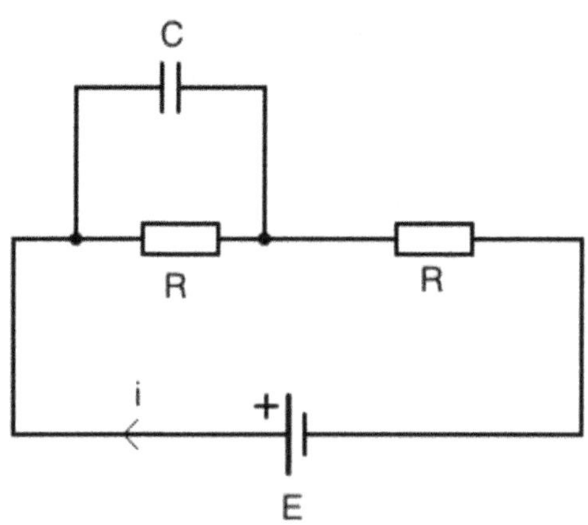

Figure M5.3.

References

[1] Newton I 1687 *Philosophiae Naturalis Principia Mathematica* (Berkeley, CA: University of California Press) (Engl. transl. A Motte)

[2] Galilei G 1953 *Dialogue Concerning the Two Chief World Systems* (Berkeley, CA: University of California Press) (Engl. transl. S Drake) pp 186–7

[3] Einstein A 1905 Zur Elektrodynamik bewegter Köpar *Ann. Phys., Lpz.* **17** 891
Einstein A 1923 On the electrodynamics of moving bodies *The Principle of Relativity* (Engl. transl. G Barker Jeffery and W Perrett) (London: Methuen)

[4] Maxwell J C 1873 *Treatise on Electricity and Magnetism* vol 1 and 2 (Clarendon Press: Oxford)

[5] Maxwell J C 1865 Dynamical Theory of the Electromagnetic Field. Part 1 *Phil. Trans. R. Soc.* **155** 459–512

[6] Drude P 1900 Zur Elektronentheorie der metalle *Ann. Phys., Lpz.* **306** 566

[7] Ohm G S 1827 *Die galvanische Kette, mathematisch bearbeitet* (Berlin) Reprinted in *Ges. Abh.* 61–186
Ohm G S *et al* 1827 *The galvanic circuit investigated mathematically* (New York: Van Nostrand) (Engl. transl. W Francis)

[8] Darrigol O 2000 *Electrodynamics from Ampère to Einstein* (Oxford: Oxford University Press) p 70

[9] Poynting J H 1884 On the transfer of energy in the electromagnetic field *Phil. Trans. R. Soc.* **175** 343–61

[10] Grant I S and Phillips W R 2008 *Electromagnetism* (*Manchester Physics Series*) 2nd edn (Hoboken, NJ: Wiley) ch 2 and 6

[11] Rosser W G V 1970 Magnitudes of surface charge distributions associated with electric current flow *Am. J. Phys.* **38** 265

[12] Galili I and Goihbarg E 2004 Energy transfer in electrical circuits: a qualitative account *Am. J. Phys.* **73** 141

Chapter 6

Magnetism

6.1 Introduction

A magnetic field is the physical phenomenon produced by moving electric charges and is exhibited by ferrous materials. At any point it is a vector quantity, i.e. it is specified by both a 'direction' and a 'magnitude'. The main attributes of magnetic field are as follows.

- A magnetic field will always be dipole, i.e. it will always have two poles, 'north' and 'south'. There are no individual magnetic poles or monopoles.
- If a magnet is divided, this will result in a smaller magnet of reduced strength with the two poles.
- Like poles in magnets will always repel each other, while poles that are opposite in nature will attract each other.
- Metallic objects block electrostatic flux, whereas magnetic flux passes through most metals with little or no effect. However, metals like iron and nickel (and alloys containing them) affect the magnetic flux by concentrating more lines of flux and are therefore known as *ferromagnetic materials*.

Hans Christian Oersted first observed the magnetic field phenomenon in 1820 [1], when a current-carrying wire was brought near a magnetic compass. The compass needle deflected at a right angle to the wire. This experiment showed that there is a relation between electricity and magnetism, but Oersted could not provide the reason for this relationship. However, his findings motivated French physicist André-Marie Ampère to develop a relationship between current-carrying conductors and the magnetic field produced from them. Many other physicists, such as Carl Friedrich Gauss, Jean-Baptiste Biot and Félix Savart, made great contributions in the area of magnetism. In 1831, Michael Faraday's findings helped to establish the relationship between magnetism and electricity. Later on, James Clark Maxwell analyzed and expanded these insights into the 20 very important Maxwell equations, thus integrating electricity and magnetism into electromagnetism. These equations form the basis of electromagnetism and were used by Einstein to establish his theory of special relativity.

It is now a well-known fact that 'there will be a magnetic field around a conductor when current flows through it'.

Thus if a snapshot of a conductor without current and a snapshot of a current-carrying conductor are the same, then as per the laws of physics the compass should show the same result when kept near the conductor in both the cases. However, the compass needle is deflected when kept near a current-carrying conductor, which means that the two snapshots cannot be same. This again verifies the fact that charges in a current-carrying conductor are Lorentz contracted (as discussed in chapter 5) and thus result in a magnetic field.

6.2 Magnetic field due to electric current

A magnetic field is the magnetic effect of electric currents and magnetic material. The intensity of a magnetic field is directly proportional to the magnitude of the current carried by the wire (discussed in section 6.3) and the direction of the magnetic field depends on the direction of the electrical current in the wire, as discussed below.

Magnetic field around a current-carrying conductor

The magnetic field lines form concentric circles around a cylindrical current-carrying conductor and the direction of such a magnetic field can be determined using the 'right-hand grip rule', as shown in figure 6.1. According to this rule, if the thumb is pointing towards the direction of the current, the fingers are pointing in the direction of the magnetic lines.

The strength of the magnetic field decreases with distance from the wire.

Magnetic field due to a current-carrying circular loop

Soon after Oersted discovered magnetic field around a current-carrying conductor, Ampère realized that twisting wire in a circular loop could increase the magnetic effect. Each small segment of the wire is surrounded by circular magnetic lines of force, as shown in figure 6.2.

The right-hand rule indicates that all the magnetic lines enter at one face and leave from the other, thus enhancing the magnetic field. The face of the coil from which the magnetic lines emerge is the north pole and the face into which they enter is the south pole. *Since the magnetic field is produced due to the flow of current and current can flow only in a closed path, this concept leads to the conclusion that electromagnets do not have magnetic monopoles. Although as per quantum mechanics some condensed matter systems contain non-isolated magnetic monopoles, magnetism in electromagnets does not emerge from monopoles.*

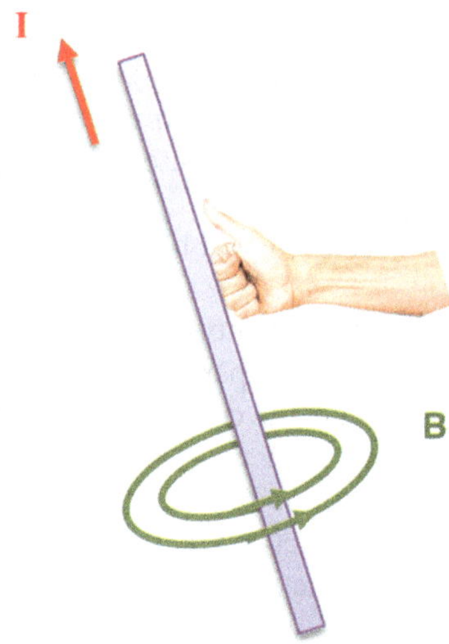

Figure 6.1. Right-hand thumb rule.

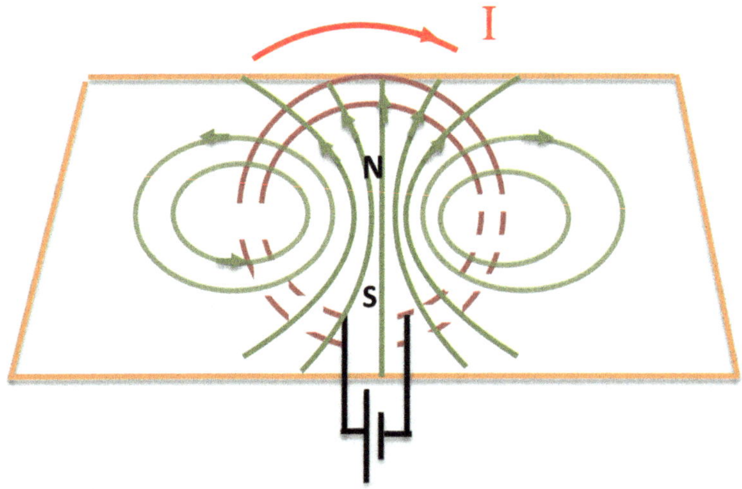

Figure 6.2. Magnetic field due to current-carrying circular loop.

Magnetic field inside a solenoid

When bent into multiple closely spaced loops, a conductor forms a coil or solenoid. A current-carrying solenoid concentrates the magnetic field inside the coil while weakening it outside. The direction of the magnetic field is estimated by wrapping the right hand around the solenoid with the fingers in the direction of the conventional current and the thumb pointing in the direction of the magnetic north pole, as shown in figure 6.3.

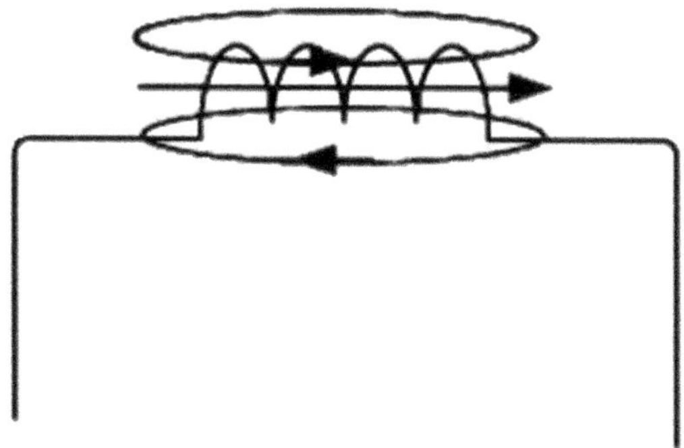

Figure 6.3. Magnetic field inside a solenoid.

6.3 Biot–Savart's law

In 1820, on the basis of their experiments, two French scientists, Jean-Baptiste Biot and Félix Savart, established a very important, fundamental and quantitative relationship between electric current and the magnetic field produced by it, which became known as the Biot–Savart law.

An electric current flowing in a conductor produces a magnetic field in the area surrounding the conductor. The magnitude of the magnetic field at a particular point in the surrounding area may be considered as the vector sum of the magnetic fields produced by each small segment of the current-carrying conductor individually. The Biot–Savart law estimates the magnitude of the magnetic field at a specific point in space due to the small segment of a current-carrying conductor. Let us evaluate magnetic field using the Biot–Savart law at a point S, which is at a perpendicular distance ($PS = R$) from an infinitely long current (I)-carrying conductor.

Consider a small element $\mathrm{d}l$ on the wire, with the line ($AS = r$) joining element $\mathrm{d}l$ to S making an angle θ with the direction of the current element, as shown in figure 6.4.

As per this law, the magnetic field $\mathrm{d}B$ at any point S due to a small current element $I{\cdot}\mathrm{d}l$ of length $\mathrm{d}l$ carrying current I is given as

$$\mathrm{d}B \propto \frac{I\mathrm{d}l\,\sin\theta}{r^2}$$

$$\mathrm{d}B = \frac{\mu_0}{4\pi}\frac{I\mathrm{d}l\,\sin\theta}{r^2}, \tag{6.1}$$

where $\mu_0 = 4\pi \times 10^{-7}$ is a magnetic constant

$$\mathrm{d}\overline{B} = \frac{\mu_0}{4\pi}\frac{I\mathrm{d}\bar{l} \times \bar{r}}{r^3}. \tag{6.2}$$

Equation (6.1) can be written as equation (6.2), where product $\mathrm{d}l \times r$ has a magnitude $\mathrm{d}l\ r\ \sin\theta$ and is directed perpendicularly to both $\mathrm{d}l$ and r, which is perpendicular to the plane of the paper and going into it, as discussed earlier, and is thus represented by a cross at S.

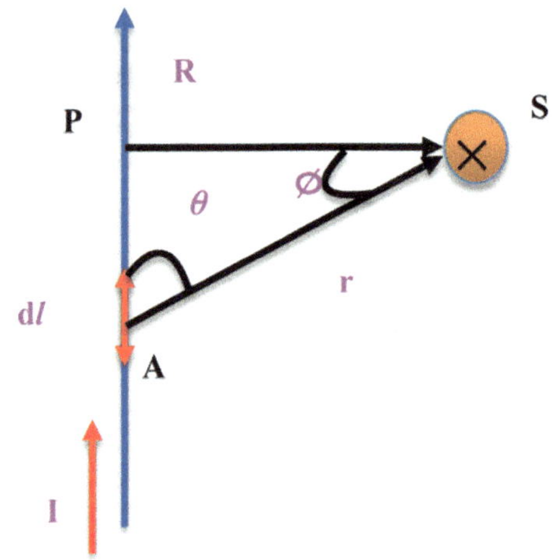

Figure 6.4. Magnetic field due to the current-carrying element.

> The Biot–Savart law concerns a fundamental quantitative relationship between an electric current and the magnetic field it produces. According to it, the value of a magnetic field produced by a current-carrying segment at a point is:
> - directly proportional to the value of current in the conductor;
> - directly proportional to the length of the current-carrying segment under consideration;
> - directly proportional to the orientation of the point with respect to the current-carrying segment;
> - inversely proportional to the square of the distance from the point of the current-carrying segment.

6.3.1 Calculation of the magnetic field due to current-carrying conductors

Now let us estimate the magnetic field for a wire of finite length l using Biot–Savart's law and equation (6.2) (figure 6.5).

From figure 6.5, $r = \dfrac{R}{\sin\theta}$.

On putting the value of r into equation (6.2), we obtain

$$B = \frac{\mu_0}{4\pi}\int\frac{\sin\theta}{r^2}\mathrm{d}l = \frac{\mu_0 I}{4\pi}\int\frac{\sin^3\theta}{R^2}\mathrm{d}l. \tag{6.3}$$

Also from figure 6.5, $l = R\tan\phi$

$$\mathrm{d}l = R\sec^2\phi\,\mathrm{d}\phi \tag{6.4}$$

$$r = R\sec\phi \tag{6.5}$$

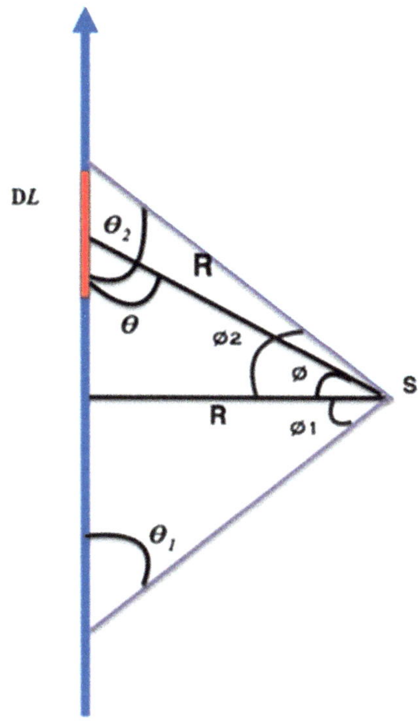

Figure 6.5. Magnetic field due to the current-carrying element of the finite element.

$$\theta = \frac{\pi}{2} - \phi. \tag{6.6}$$

On putting the value of dl, r and θ from equations (6.4)–(6.6) into equation (6.3), we obtain

$$\overline{B} = \frac{\mu_0}{4\pi}\int \frac{IR\sec^2\phi \, \mathrm{d}\phi \sin\left(\frac{\pi}{2} - \phi\right)}{R^2\sec^2\phi} = \frac{\mu_0}{4\pi}\int \frac{I\cos\phi \, \mathrm{d}\phi}{R}.$$

On integrating this between angle ϕ_1 and ϕ_2, we obtain

$$B = \frac{\mu_0 I}{4\pi R}\int_{\phi_1}^{\phi_2} \cos\phi \, \mathrm{d}\phi$$

$$B = \frac{\mu_0 I}{4\pi R}\left(\sin\phi_2 - \sin\phi_1\right).$$

On taking a negative sign ϕ_1 in order to consider the portion of the length extended in the negative y-axis from the origin,

$$B = \frac{\mu_0 I}{4\pi R}\left(\sin\phi_2 + \sin\phi_1\right).$$

On putting the value of ϕ_1 and ϕ_2 into the above equation, we obtain

$$B = \frac{\mu_0 I}{4\pi R}\left[\sin\left(\frac{\pi}{2} - \theta_1\right)\right] + \sin\left(\frac{\pi}{2} - \theta_2\right)$$
$$B = \frac{\mu_0}{4\pi R}\left[\cos\theta_1 + \cos\theta_2\right]. \tag{6.7}$$

Special cases

(i) If the field point S is located along the perpendicular bisector, then $\theta_1 = \theta_2$ and $\cos\theta_1 = \frac{L/2}{\sqrt{\left(\frac{L}{2}\right)^2 + R^2}}$. The magnetic field is given by equation (6.8):

$$B = \frac{\mu_0 I}{2\pi R}\cos\theta_1 = \frac{\mu_0 I}{2\pi R}\frac{L/2}{\sqrt{\left(\frac{L}{2}\right)^2 + R^2}}. \tag{6.8}$$

(ii) If the length of the wire is infinite, then the magnetic field at a distance R from the wire is given by equation (6.9):

$$\overline{B} = \frac{\mu_0 I}{2\pi R}. \tag{6.9}$$

6.4 Ampère's law

In 1826, André-Marie Ampère developed a relation between magnetic fields and electric currents. Ampère's law says precisely that for any closed path, the line integral of the magnetic field is proportional to the electric current that passes through that enclosed path with a constant of proportionality equal to the permeability of free space, as represented by equation (6.10):

$$\oint \overline{B} \cdot \mathrm{d}\overline{l} = \mu_0 I_{\text{enc}} \tag{6.10}$$

In order to obtain an insight into Ampère's law, let us consider an infinitely long thin wire that is carrying a steady current I, with a closed loop C encircling this wire in an anti-clockwise direction according to the right-hand grip rule discussed earlier. If Δl (as shown in figure 6.6) is a small straight segment of this loop then, as per Ampère's law, the sum of the length elements times the magnetic field in the direction of the length element is equal to the permeability times the electric current enclosed in the loop, as illustrated by equation (6.11):

$$\Sigma B_1\,\Delta L = \mu_0 I. \tag{6.11}$$

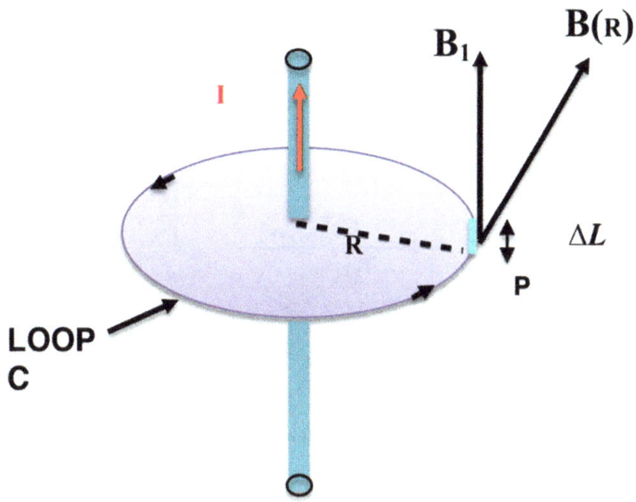

Figure 6.6. Ampère's law.

This field is orthoradial with $\overline{B} = B(R)\overline{u}(\theta)$
or

$$\oint \overline{B} \cdot \mathrm{d}\overline{l} = \mu_0 I$$
$$B(R)(2\pi R) = \mu_0 I \tag{6.12}$$
$$\overline{B}(P) = \frac{\mu_0 I}{2\pi R}\,\overline{u}(\theta).$$

Surprisingly, equation (6.12) of the magnetic field is independent of the wire's length and hence according to Ampère's law the value of the magnetic field remains the same as given by equation (6.12) for a finite wire of length l. This result is different from the result obtained using Biot–Savart's law and appears to be incorrect.

However, when $R \ll L$ the equations derived from Biot–Savart's and Ampère's laws are same.

This shows that Ampère's law is only applicable only when current flows through the circuit. Current cannot flow through a finite wire, because the wire is not closed, whereas in the case of an infinite wire one can say that the wire is closed at infinity.

Thus Ampère's law as given by equation (6.10) is only applicable to an infinite wire and it is thus not a generalized equation; it was therefore re-derived by James Clerk Maxwell in 1861.

This modified equation, known as the Maxwell–Ampère equation, is discussed in chapter 8 after the introduction of displacement current.

6.5 Magnetic forces

Moving electrically charged particles exert the force of attraction or repulsion on each other, depending on the polarity of their charge. This force (known as magnetic

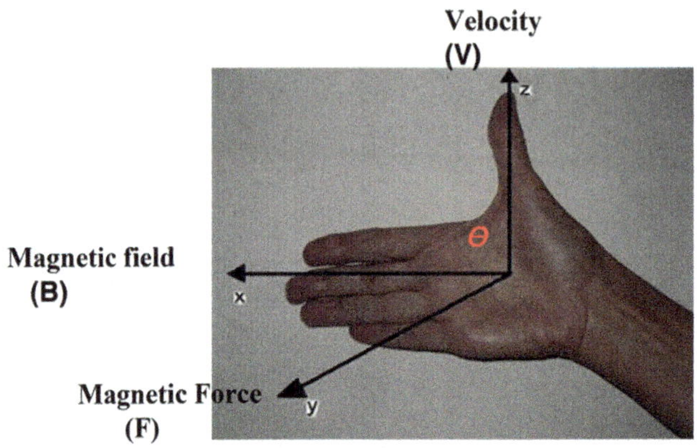

Figure 6.7. The right-hand rule for determining the direction of magnetic force.

force) can be explained as the effect exerted upon either charge by a magnetic field produced by the other charge.

The magnitude of a magnetic force on a charge q moving at a velocity v in a magnetic field of magnitude B is given by equation (6.13):

$$F = qvB \sin \theta, \tag{6.13}$$

where θ is the angle between the direction of the magnetic field (B) and the direction of the velocity (v) of q. Thus the magnetic force on the charge is maximal when a charge moves perpendicularly to the magnetic field and zero when the charge is either stationary or moving in the direction of the magnetic field. The direction of this magnetic force on a charge is perpendicular to the plane formed by the direction of the velocity of that charge and the direction of the magnetic field surrounding it as determined by the right-hand rule. According to the right-hand rule, the direction of the magnetic force on a positive charge is determined by placing the thumb of the right hand in the direction of the velocity of that charge, with the fingers in the direction of the magnetic field, as shown in figure 6.7. The perpendicular to the palm indicates the direction of the force. However, for a negative charge the force is in the opposite direction to that for a positive charge.

We know that electric force exists between the charges irrespective of whether they are stationary or moving, and it is thus referred to as *electrostatic* force. Hence if a charged particle moves in a region with both electric and magnetic fields, it experiences the combination of electric and magnetic forces known as Lorentz force, as given by equation (6.14):

$$\overline{F} = q\left(\overline{E} + \overline{v} \times \overline{B}\right). \tag{6.14}$$

- Stationary charged particles create electric field in their surrounding area.
- The electric field exerts an electric force ($F = qE$) on other charges present in the field.

- A moving charge or current produces a magnetic field along with an electric field in the surrounding area.
- The magnetic field exerts a magnetic force ($\overline{F}_m = q(\overline{v} \times \overline{B})$) on other moving charges or current present in that field.
- If the particle's charge is zero, then the electric and magnetic fields will have no effect on it.
- If a particle's velocity is zero, then the magnetic field has no effect.
- If the particle is moving in the same direction as the magnetic field, then no magnetic force is felt.
- If a charged particle is moving in a given direction in the presence of magnetic field, then the particle will accelerate at right angles to both the magnetic field and its initial direction. In other words, magnetic fields can never do any work on a charged particle.

6.6 Electric and magnetic fields: consequences and genesis

The above discussion leads us to the conclusion that when an electric field is applied to a conductor, it gives rise to a current that in turn produces a magnetic field. This indicates that a magnetic field is produced by an electric field, but raises the question of whether an electric field is produced by a magnetic field.

In 1831, Michael Faraday came up with the answer to the above question while conducting experiments in his lab. Faraday's experimental demonstration of electromagnetic induction established that an electric current is induced in the coil when it is subjected to a varying magnetic field. Thus, Faraday's law states that *when the magnetic flux linking a circuit changes, an electromotive force is induced in the circuit proportional to the rate of change of the flux linkage*, as expressed by equation (6.15):

$$E \propto \frac{\mathrm{d}\phi_{\mathrm{B}}}{\mathrm{d}t}. \tag{6.15}$$

where E is the emf and ϕ_{B} is the magnetic flux.

The magnetic flux ϕ_{B} can be varied by varying the strength of the magnetic field, moving the coil away from or towards the magnetic field, moving a magnet away from or towards the coil, or rotating the coil relative to the magnet.

Faraday's law gives the magnitude of the induced emf, but does not say anything about the direction of this induced emf. In 1834, the German scientist Heinrich Lenz formulated Lenz's law stating that

> *If an induced current flows, its direction is always such that it opposes the change that produces it.*

Thus Faraday's law in combination with Lenz's law states that

> *When an emf is produced by a change in the magnetic flux, the polarity of the induced emf is such that it produces a current whose magnetic field opposes the change that produces it.*

$$E = -\frac{\mathrm{d}\phi}{\mathrm{d}t}.$$

As such, the findings of the physicists H C Oersted and Michael Faraday corrected the myth that electricity and magnetism are two independent phenomena and made the world realize that electric field and magnetic field are two parts of a greater whole: *the electromagnetic field.*

6.7 Magnetism as a relativistic effect

Static electromagnetic fields (not varying in time) have both electric and magnetic components, but for a few observers only one of the fields exists, as for them the other field is suppressed. However, the suppressed field appears to be dominant in some other observer's frame.

In 1953 Albert Einstein wrote 'What led me more or less directly to the special theory of relativity was the conviction that the electromotive force acting on a body in motion in a magnetic field was nothing else but an electric field [2].'

Hence according to the special theory of relativity, the splitting of electromagnetic force into electric and magnetic components is not fundamental, but varies with the observer's frame of reference, i.e. 'electric and magnetic fields transform depending on various inertial frames of reference'.

Qualitative explanation
The above statement can be explained with the help of an elevator thought experiment. Let us consider two inertial frames of reference, S_1 and S_2. Frame S_2 is at rest with respect to the ground and frame S_1 is an elevator that is moving with a constant velocity v, as shown in figure 6.8.

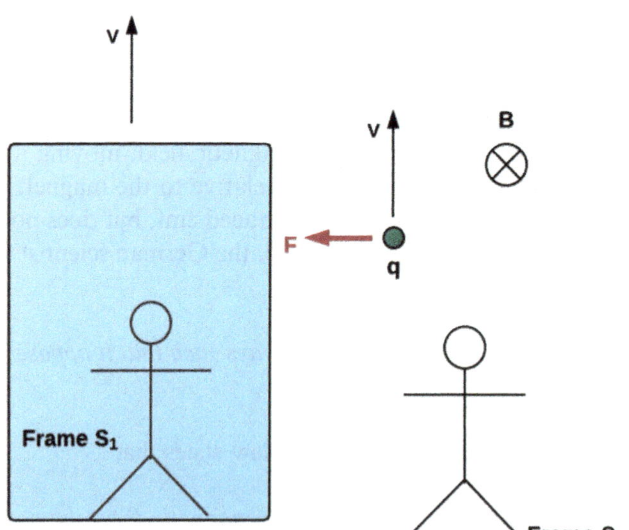

Figure 6.8. Elevator thought experiment.

Let us assume a uniform magnetic field of magnitude B into the page outside the elevator. There is a point charge particle of charge q outside the elevator that is moving with the same constant velocity v in the direction shown. In frame S_2, the observer says that there should be force on the point charge particle towards the left due to magnetic field, as dictated by the right-hand rule. However, with respect to the observer in frame S_1, the charge is at rest and hence the magnetic field should not exert any force on the point charge particle. But the principle of relativity teaches us that there is no special inertial frame of reference in which a physical phenomenon takes place. If the charge particle experiences a force towards the left according to the observer in frame S_2, it should experience the same amount of force as observed from frame S_1. Since forces are transformed between different inertial frames of reference, the final effect should be the same for both observers. Hence, the magnetic field into the page as seen by observer in frame S_1 should become an electric field directed towards the left.

Thus, a magnetic field is transformed into an electric field when viewed from a different frame of reference. Hence, electric and magnetic fields are not different entities, and are aspects of a much broader entity called electromagnetic field.

Quantitative explanation

The present definition of current in a metallic conductor is based on the Drude theory of electric conduction. A simple model of an infinitely long current-carrying conductor is considered in which positive ions are at rest and the conduction electrons move with drift velocity v_d. If the current in the conductor is i, the drift velocity is given by

$$v_d = \frac{i}{neA}, \tag{6.16}$$

where A is the area of cross-section of the conductor, e is the charge of an electron and n is number of free electrons per unit volume of the conductor. According to Ohm's law, the current i flowing through a metallic wire of resistance R is proportional to the potential drop V across the wire. This is given by

$$i = \frac{V}{R}. \tag{6.17}$$

We can immediately see a problem with equation (6.16). In principle, according to equation (6.17), the value of the current i flowing through a conductor can be made as large as possible by decreasing the resistance of the conductor between which a constant potential difference is applied. Since the denominator in equation (6.16) is a constant, the drift velocity can also be made as large as possible, even greater than the speed of light. This result is inconsistent with the special theory of relativity. Hence, we need to define the terms used in equation (6.16) more carefully.

We shall now present a possible solution to the problem above, which was also discussed in chapter 5.

It is a well-known fact that if charges are in motion, their volume density changes because of Lorentz's contraction. The volume density changes as the time component of a four-vector. In particular the four-vector $[\rho c, \vec{j}\,]$, with ρ being the charge density, **J** being the current density and c being the speed of light, transforms according to Lorentz's transformation. According to the available literature, this Lorentz contraction in a metallic current-carrying conductor is only considered when there is relative motion between the charge and the conductor. However, Lorentz's contraction not only exists when there is relative motion between the charge and the conductor, but also when the current flows through the conductor. We are then led to a new definition of current, which is illustrated as follows.

If the current in the conductor is i, the drift velocity is given by equation (6.16). We define n as the number of free electrons per unit volume of the conductor when the conductor is at rest. Since n changes as a time component of a four-vector, it can also be written as

$$n = n^{\dagger}\gamma, \tag{6.18}$$

where $n^{\dagger}$ represents the number of free electrons per unit volume when the conductor is at rest and does not carry any current, and $\gamma = \dfrac{1}{\sqrt{1 - \dfrac{v_d^2}{c^2}}}$.

We can then write equation (6.16) as

$$i = \lambda^{\dagger} v_d, \tag{6.19}$$

where $\lambda^{\dagger} = neA$, which is a constant for a given conductor. $\lambda^{\dagger}$ also represents the electron charge per unit length of the conductor. Let λ represent the charge per unit length of the electrons when there is no current through it. Then we have

$$\lambda^{\dagger} = \lambda\gamma. \tag{6.20}$$

This can be for both positive and negative charges in the conductor, where λ_+ is taken to be the charge per unit length of positive ions and λ_- is taken to be the charge per unit length of electrons. Further, we assume that the charge densities are equal for a conductor carrying no current through it. Hence

$$\lambda_+ = \lambda_- = \lambda.$$

Thus, the point charged particle experiences no first- or second-order forces when it is kept near a conductor that carries no current through it. We can write our new definition of current as

$$i = \frac{n^{\dagger} e A v_d}{\sqrt{1 - \dfrac{v_d^2}{c^2}}}. \tag{6.21}$$

With this definition, we can see that the drift velocity of the electrons cannot be greater than c. However, the value of the current can be as large as possible. We shall provide a second motivation to redefine current, as shown in equation (6.21) later in this section.

Let the current be i. The electrons are moving with a drift velocity v_d in a particular direction. Consider one frame S, which is at rest with respect to the charged particle with a charge q. The charged particle is also at rest with respect to the conductor.

In frame S,

$$\lambda'_+ = \lambda$$

$$\lambda'_- = \frac{\lambda_-}{\sqrt{1 - \frac{v_d^2}{c^2}}} = \frac{\lambda}{\sqrt{1 - \frac{v_d^2}{c^2}}}.$$

The primed values of λ indicates the charge per unit length as viewed from frame S. Thus, the charges no longer balance and a net charge is developed. This developed net charge per unit length is given by

$$\lambda' = \lambda_+ - \lambda_- = \lambda[1 - \gamma],$$

where $\gamma = \frac{1}{\sqrt{1 - \frac{v_d^2}{c^2}}}$.

Thus, as $\gamma > 1$, a negative electric field is produced radially because of this inequality of the positive and negative charge distributions. The strength of this field, at a distance r from the conductor, is given by:

$$E = \frac{\lambda'}{2\pi\varepsilon_0 r} = \frac{\lambda[1 - \gamma]}{2\pi\varepsilon_0 r},$$

where ε_0 represents the permittivity of free space.

Accordingly, the force on the charged particle with charge q due to this field is

$$F = q\frac{\lambda[1 - \gamma]}{2\pi\varepsilon_0 r}. \tag{6.22}$$

To show that the force given by equation (6.22) is indeed a second-order force in drift velocity v_d to the first approximation, we write it as

$$F \approx \frac{\lambda q}{2\pi\varepsilon_0 r}\left[1 - \left(1 + \frac{v_d^2}{2c^2}\right)\right] \tag{6.23}$$

using the approximate binomial expansion and the fact that practically, $v_d \ll c$. In fact, it is now known that $v^2/c^2 < 10^{-20}$ for essentially all cases for metallic conductors at room temperature. Thus, equation (6.23) is a valid approximation of equation (6.22).

Equation (6.23) can then be reduced to

$$F \approx \frac{\lambda q v_d^2}{4\pi\varepsilon_0 r c^2}. \tag{6.24}$$

Equation (6.24) shows that the current-carrying conductor produces a small amount of electric field pointing radially toward the wire that is of second order to the first approximation.

Now, we shall present a complete derivation of the magnetic field produced in a current-carrying conductor as a relativistic effect.

Let the velocity of the charged particle be u with respect to the conductor. We assume here, for simplicity, that the charged particle moves to the left (negative direction) and that the current i has the opposite direction, i.e. it moves to the right (positive direction), see figures 6.9 and 6.10. Consider a frame S that is at rest with respect to the charged particle and frame $S^{\dagger}$, which is at rest with respect to the conductor.

From frame S, the positive ions move with velocity u in the positive direction and the conduction electrons also move with velocity v_{-} in the same direction, given by

$$v_{-} = \frac{u - v_d}{1 - \frac{u v_d}{c^2}}. \tag{6.25}$$

At this point, we define $\gamma_1 = \frac{1}{\sqrt{1 - \frac{u^2}{c^2}}}$.

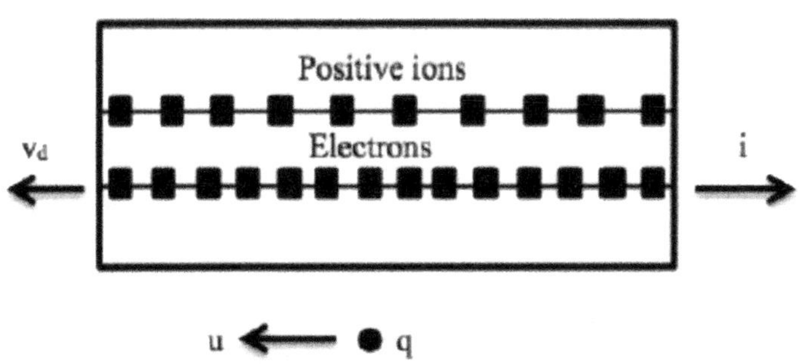

Figure 6.9. Conductor as seen from its own rest frame $S^{\dagger}$.

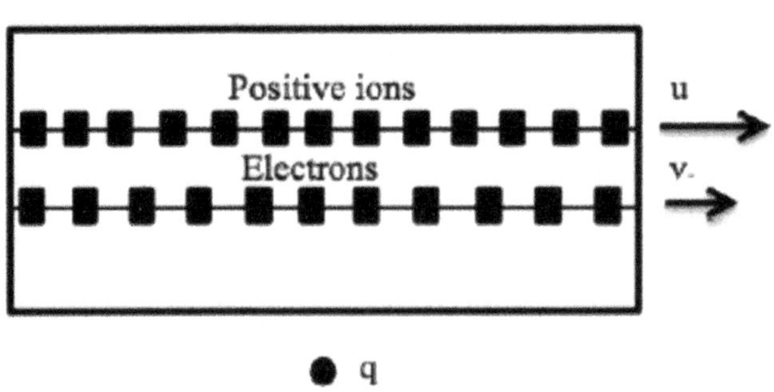

Figure 6.10. Conductor as seen from frame S.

As positive ions move with a velocity u with respect to charge, the charge distribution changes and is given by

$$\lambda_+ = \lambda\gamma_1.$$

And similarly,

$$\lambda_- = \frac{\lambda}{\sqrt{1 - \dfrac{v_-^2}{c^2}}}. \tag{6.26}$$

Substituting v_- from equation (6.25) into equation (6.26), we obtain

$$\lambda_- = \lambda\gamma\gamma_1\left(1 - \frac{uv_d}{c^2}\right).$$

Thus, the total charge density as seen from frame S is:

$$\lambda' = \lambda_+ - \lambda_- = \lambda\gamma_1\left(1 - \gamma + \frac{uv_d\gamma}{c^2}\right).$$

The force on the charged particle with charge q placed at a distance r from the conductor is given by

$$F = \frac{q\lambda\gamma_1\left(1 - \gamma + \dfrac{uv_d\gamma}{c^2}\right)}{2\pi\varepsilon_0 r}. \tag{6.27}$$

However, we need the force F_0 in the rest frame of the conductor (frame $S^\dagger$) so as to relate it to a familiar form. The force F in frame S is to be converted to frame $S^\dagger$ by

$$F_0 = \frac{F}{\gamma_1}.$$

Thus, from equation (6.27), the force on the particle is

$$F_0 = \frac{q\lambda\left(1 - \gamma + \dfrac{uv_d\gamma}{c^2}\right)}{2\pi\varepsilon_0 r}. \tag{6.28}$$

With the permeability of space denoted by μ, the speed of light is, according to Maxwell, given as

$$c = \frac{1}{\sqrt{\varepsilon_0\mu_0}}.$$

Substituting the above formula into equation (6.28), we obtain

$$F_0 = Buq + \frac{q\lambda(1 - \gamma)}{2\pi\varepsilon_0 r}, \tag{6.29}$$

where $B = \dfrac{\mu i}{2\mu r}$ is the known formula used to calculate the strength of the magnetic field at a distance r from the conductor. Note that we used the new definition of current defined previously in equation (6.21) in equation (6.29). The second motivation to do this is the fact that the Lorentz force form, as given by equation (6.14), is relativistically correct, and to preserve its form we should redefine current.

Equation (6.29) shows that when a current flows through a conductor, not only is a magnetic field created, but so is a second-order (to first approximation) electric field. The first and second terms represent the magnetic and electric fields produced, respectively. Equation (6.29) can be represented in vector form as follows

$$\vec{F_0} = q\left[\left(u \times \vec{B}\right) + \vec{E}\left(1 - \gamma\right)\right], \tag{6.30}$$

where $E = \dfrac{\lambda}{2\pi\varepsilon_0 r}$.

We can also write equation (6.30) in Lorentz form as

$$\vec{F_0} = q\left[\left(u \times \vec{B}\right) + \vec{E}^{\dagger}\right],$$

where, $\vec{E}^{\dagger} = \vec{E}(1 - \gamma)$.

Thus, the electric field is created because of the charge imbalance between the positive ions and electrons caused by the flow of the current together with the magnetic field.

Hence we can conclude that *electric and magnetic fields are not different entities. They are aspects of a much broader entity called electromagnetic field.*

Further, from the Oersted experiment it was clear that magnetic field is produced because of a current-carrying conductor, but it was not clear that current is produced due to moving charges, or in other words one can say that the moving charges create the same effect as current [1]. This question was put to rest in the year 1878 by Henry A Rowland when he demonstrated a rotating disc experiment, as discussed in the next section.

6.8 Rowland's experiment

Henry Rowland demonstrated that magnetic field is produced by moving charges and thus made it clear that an electric current is nothing but the movement of charges in matter.

In order to substantiate this he prepared a slotted disk of insulating material with a layer of gold deposited on the surface. This disc was rotated at a frequency of 50 rev s^{-1} around the shaft inside a cylindrical narrow case with a small air gap, as shown in figure 6.11. The disc was charged by applying a voltage between the disc and the case. The slit in the disc ensured that the charges could not be in conducting motion and were only moved when the disc rotated. On the rotation of the disc, a small magnetic field was detected by a magnetic compass suspended at a nearby location.

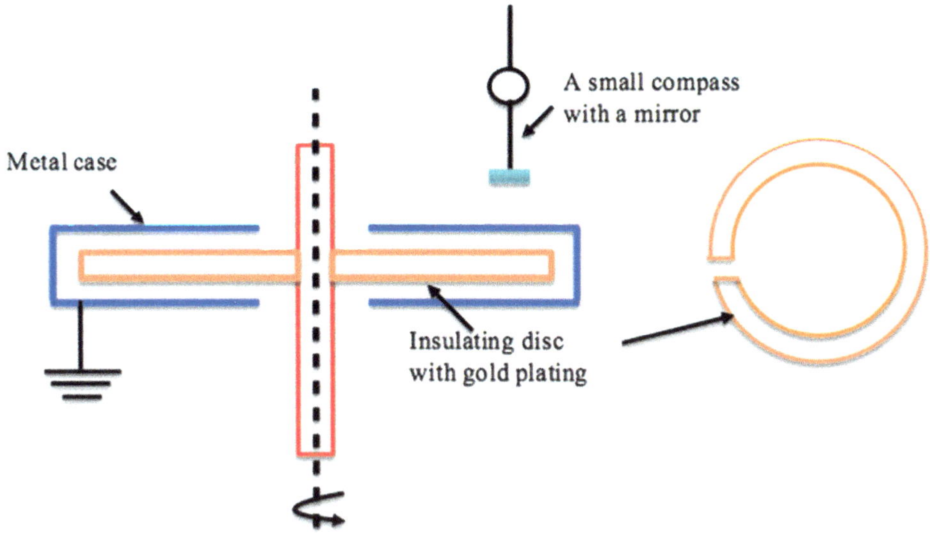

Figure 6.11. Rowland's experimental setup.

Thus moving charges created the same effect as current, leading to a very important conclusion, which seems obvious now, that *an electric current is due to movement of charges.*

A very striking fact about this experiment was that it did not reveal anything about the polarity of the moving charge. That is, one could not say whether an electric current was due to the movement of negative or positive charges [3].

This unaddressed issue from Rowland's experiment led to the discovery of the Hall effect by Edwin Herbert Hall.

6.9 The Hall effect

It is a well-known fact that the mobile charges in a conventional conducting material are negatively charged, i.e. electrons. The question is, how do we justify the above statement? The Hall effect, which was discovered by Edwin Herbert Hall in 1879, provides experimental evidence for the fact that electrons are the mobile charges in any conducting material [4].

In order to understand the effect, let us place a thin flat metallic ribbon perpendicular to a uniform magnetic field B, as shown in figure 6.12(*a*). If the current I is passed along the length of the conductor, there are two possibilities, either the current is carried by the protons moving from left to right, or by the electrons moving from right to left, as shown in figure 6.12 (*b*).

If the movement of protons from left to right carries the current, then a force will be exerted by the magnetic field on these protons. Due to this force, the protons will be deflected in the upward direction, which will result in the upper edge being positively charged, while the lower edge will become negatively charged. Consequently, there will be positive potential difference V_{H} across the upper and lower edges of the conducting material.

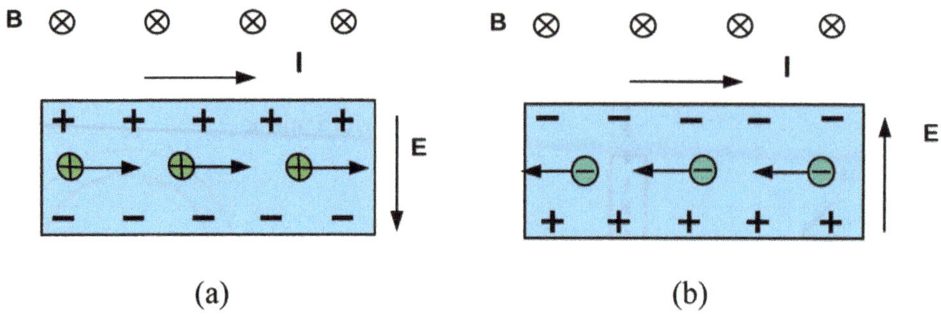

Figure 6.12. The Hall effect for (a) positive carriers and (b) negative carriers.

Now if we assume that the current is carried by electrons from right to left, then these particles will also be deflected in the upward direction by the magnetic field. This should result in the upper and lower edges being negatively and positively charged, respectively. The potential difference will now be negative.

The potential difference is called the Hall voltage. If the Hall voltage is positive, the moving charges are protons, and if the Hall voltage is negative, the moving charges are electrons. Thus from the Hall voltage one can predict the polarity of the moving charges.

When this experiment was conducted with a metal strip, it was observed that the upper edge is always negative and the lower edge is always positive. Hence it was concluded that in the case of metals *negatively charged electrons are the current carriers.*

Once the direction of the Hall voltage (V_{H}) is known, it is important to estimate the magnitude of it. Let us consider a negative charge q moving perpendicularly to the magnetic field along the metal ribbon with the drift velocity v_d, as shown in figure 6.12(b). As the charge is moving perpendicularly to the magnetic field, a magnetic force ($F_m = qv_dB$) is exerted on the mobile charge in the upward direction. This results in the accumulation of negative charges on the upper face and positive charges on the lower face. The separation of charges establishes an electric field (E) from the upper to the lower face of the ribbon and a steady electrical potential. If the electrical potential is V_{H} and the width of the ribbon is w, then the magnitude of the electric field is given by equation (6.31):

$$E = \frac{V_{\text{H}}}{w}. \tag{6.31}$$

This electric field opposes the migration of further charges and exerts electric force ($F_e = qE$) on the charge in the direction just opposite to that of magnetic force. At steady state, electric force is equal to magnetic force:

$$\begin{aligned} F_e &= F_m \\ qE &= qv_dB. \end{aligned} \tag{6.32}$$

On putting the value of E from equation (6.31) into equation (6.32), we obtain

$$\begin{aligned} q\frac{V_H}{w} &= qv_dB \\ V_H &= v_dBw. \end{aligned} \quad (6.33)$$

From equation (6.33), it is clear that the Hall voltage V_H is directly proportional to the magnitude of the magnetic field B.

6.10 The energy associated with the magnetic fields

It is universally known that 'energy can neither be created nor destroyed'. It is in fact converted into another form, as resistors convert electrical energy into thermal energy and capacitors convert input energy into an electric field wherein charged particles attain potential energy. Even if the capacitor is disconnected from the input, it stores the energy and thus can be reused as a source.

A pure inductor converts electrical energy into a magnetic field, which carries the energy.

When an inductor is connected across a battery, a magnetic field is created that carries electrical energy. However, we cannot simply disconnect the inductor from the battery and utilize its stored energy, as we do in the case of a capacitor. The reason for this is that the current flowing through an inductor should be constant in order to retain the magnetic field. Thus one needs to short-circuit the clamps of an inductor simultaneously with the disconnection of the inductor from the battery. First of all, it is not possible to do so, and then even if this were done, the current flowing through the wire of an inductor would be converted into thermal energy because of the ohmic resistance.

Conclusion

In this chapter, we have seen how a current-carrying conductor produces magnetic field. We have also seen how to calculate the magnetic field in the vicinity of a conductor using Biot–Savart and Ampère's laws. We presented Faraday's law and saw how energy is stored in a magnetic field. Using the Hall effect, we also proved that the motion of electrons in a metallic conductor constitutes the current.

Exercises

MCQs

1) A charged particle moves in a gravity-free space without change in velocity in an electric field E and a magnetic field B. Which of the following is not possible?
   a) $E = 0,\ B = 0$
   b) $E = 0,\ B \neq 0$
   c) $E \neq 0,\ B = 0$
   d) $E \neq 0,\ B \neq 0$
2) Consider the situation shown in figure M6.1. The straight current-carrying wire closed at infinity is fixed, while the loop can move freely. The loop will:

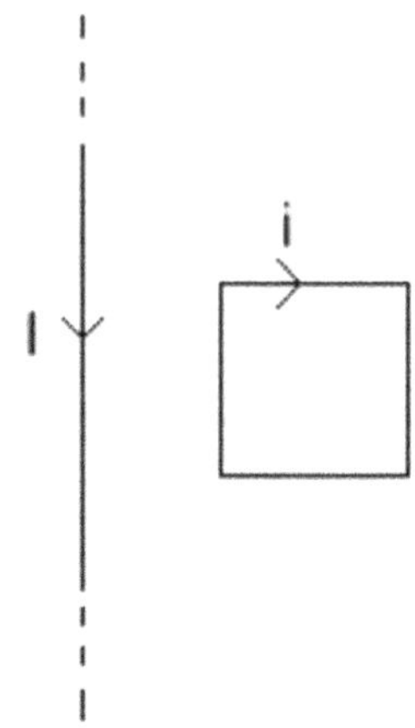

Figure M6.1.

a) Remain stationary
b) Move toward the wire
c) Move away from the wire
d) None of the above

3) A long straight wire carries a current along the Z-axis. One cannot find two points in the XY plane such that:
   a) The directions of the magnetic fields are the same
   b) The magnetic fields are equal
   c) The field at one point is opposite to that at the other point
   d) The magnitudes of the magnetic fields are equal
4) A vertical wire carries a current in the upward direction along the positive Z-axis. A proton beam sent horizontally towards the wire will be deflected:
   a) Towards the negative Z-axis
   b) Towards the positive Z-axis
   c) Towards the positive Y-axis
   d) None of the above
5) Two infinitely long current-carrying wires carrying currents i_1 and i_2 are placed parallel to each other. Suppose that when the currents are in the same direction, the magnetic field created midway between the wires is k_1. When the currents are in opposite directions, suppose that the magnetic field created at the midpoint is k_2. The ratio i_1/i_2 is equal to:
   a) k_1/k_2
   b) k_2/k_1
   c) $(k_1 + k_2)/(k_2 - k_1)$
   d) $(k_1 + k_2)/k_1$

Problems

1) A charge of 2 nC moves with a speed of 200 m s^{-1} along the positive X-axis. A magnetic field $\vec{B} = (\hat{j} + \hat{k})T$ exists in space. Find the magnetic force on the charge.

2) A long solenoid is formed by winding 40 turns cm s^{-1}. What current is necessary to produce a magnetic field of 10 mT inside the solenoid?
3) Two long straight wires, each carrying a current of 10 A are kept parallel to each other at a distance of 5 cm. Find the magnitude of the magnetic force experienced by 10 cm of a wire.
4) Suppose a magnetic field B exists in space such that it is directed into the plane of this paper. A particle of mass m is projected with a speed v in the plane of paper. The velocity is perpendicular to the magnetic field. Find the revolution frequency of the particle.
5) A proton with charge e moves with a constant speed v along a circular path of radius r. Find the magnetic moment of the circulating proton.

References

[1] Oersted H C 1820 Experiments on the effect of a current of electricity on the magnetic needles *Ann. Phil.* **16** 273

[2] Shankland R S 1964 Michelson–Morley experiment *Am. J. Phys.* **32** 16–81

[3] Purcell E M 1985 *Electricity and Magnetism (Berkeley Physics Course* vol 2*)* 2nd edn (New York: McGraw-Hill)

[4] Hall E 1879 On a new action of the magnet on electric currents *Am. J. Math.* **2** 287–92

IOP Publishing

The Foundations of Electric Circuit Theory

N R Sree Harsha, Anupama Prakash and D P Kothari

Chapter 7

Electromagnetic induction

Electricity and magnetism were considered to be unrelated phenomena for years. However, in 1820 the Oersted experiment showed that moving electric charges produce magnetic field, although it was still not clear whether magnetic field produces electric current.

It was in 1830, when Michael Faraday discovered deflection in a galvanometer while moving a permanent magnet in and out of a coil, that it was established that a voltage is generated across a conductor if it is subjected to a varying perpendicular magnetic field.

Thus an experiment demonstrated by Faraday led to a very important law, 'Faraday's law of electromagnetic induction', which states that electric current in a circuit can be produced by the force of a magnetic field instead of batteries.

This phenomenon of electromagnetics was not only important from an academic point of view, it was also important from a practical perspective, as it led to the invention of modern generators and transformers.

The Oersted and Faraday experiments led to the conclusion that these two entities are related to each other.

7.1 Faraday's experiments

Faraday's experiments are divided into three categories, as follows.

Experiment 1. Relative motion between a magnet and coil

(i) A coil C is connected to a galvanometer G with a bar magnet moving back and forth inside the coil, as shown in figure 7.1.

Observations: when the north pole of a magnet is pushed towards a coil, the galvanometer shows a deflection in one direction, and when it is pulled out it shows a deflection in the opposite direction. Similarly, if the south pole of a magnet is pushed towards the coil, a deflection in the opposite direction to that of the north pole is observed.

doi:10.1088/978-0-7503-1266-0ch7

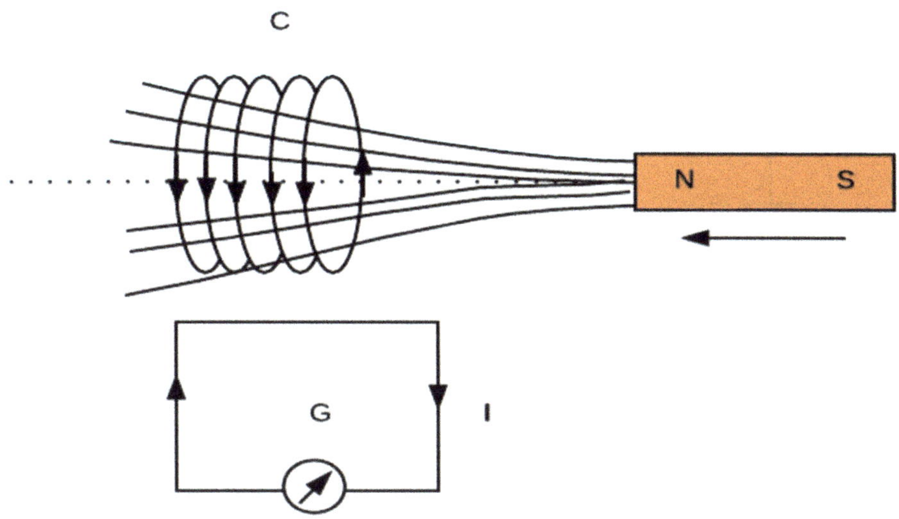

Figure 7.1. A stationary coil and a moving magnet.

(ii) When the bar magnet is stationary and the coil is moved towards or away from the magnet, similar results are obtained.
(iii) When the magnet and coil are stationary, there is no deflection in the galvanometer.
(iv) When the magnet is moved more quickly, the deflection in the galvanometer is larger.

Conclusion: *the relative motion between the magnet and the coil is the cause of electric current in the coil.*

Experiment 2. Relative motion between two coils

(i) Let us consider two coils *A* and *B* placed next to each other. A battery is connected across coil *A* and a galvanometer is connected across coil *B*, as shown in figure 7.2. When coil *B* moves toward coil *A*, a deflection is observed in the galvanometer. If coil *B* moves away from coil *A*, a deflection in the opposite direction to that of the previous one is observed.
(ii) After changing the polarity of the battery, the movement of coil *B* towards coil *A* produces a deflection in the opposite direction to that of the previous case.
(iii) If coil *B* is held stationary and coil *A* is moved towards or away from coil *B*, similar results are obtained.

Conclusion: *the magnetic flux surrounding the current-carrying coil changes due to the relative motion between it and another coil and thus causes the flow of electric current.*

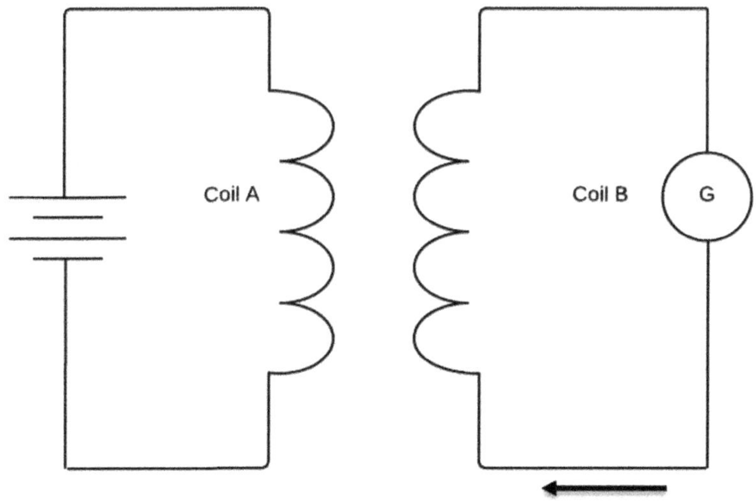

Figure 7.2. Two coils: coil A is stationary and coil B is in motion.

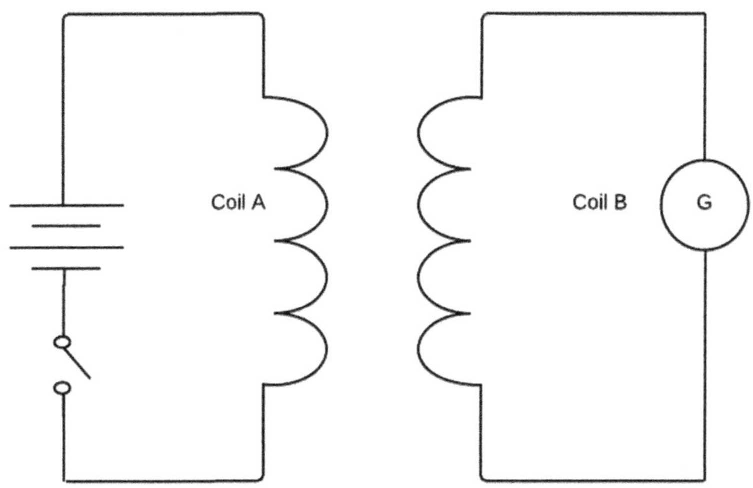

Figure 7.3. Two stationary coils and a tapping key.

Experiment 3. Two stationary coils and a tapping key

(i) Let us consider two coils A and B. Coil A is connected across the battery through a tapping key K and coil B is connected across a galvanometer, as shown in figure 7.3.

As soon as the tapping key is closed there is a momentary deflection in the galvanometer and the pointer comes back to its original state. On opening the tapping key, a momentary deflection is again observed in the galvanometer, but in the opposite direction.

Conclusion: *even though the coils are stationary, the magnetic flux changes due to a change in the position of the tapping key. When the switch is ON, current builds up in coil A from zero, and when the switch is OFF, the current reduces to zero. This momentary change in current causes a momentary change in the magnetic flux, resulting in the momentary deflection of the galvanometer.*

7.2 Faraday's law of electromagnetic induction

After analyzing the results of the above experiments, Michael Faraday came up with the law of electromagnetic induction, which states that

> *A voltage is induced in a circuit whenever relative motion exists between a conductor and a magnetic field and that the magnitude of this voltage is proportional to the rate of change of the flux.*

$$e = N\frac{\mathrm{d}\varphi}{\mathrm{d}t}, \tag{7.1}$$

where e = instantaneous induced voltage in volts, N = number of turns in a coil, φ = magnetic flux in webers and t = time in seconds.

This induced voltage is directly proportional to:

(a) the number of turns in the coil
(b) the speed of the relative motion between the coil and the conductor
(c) the strength of the magnetic field

7.3 Lenz's law of electromagnetic induction

Faraday's law gives the magnitude of a voltage (equation (7.1)) induced in a conductor when it passes through a magnetic field or a magnetic field is made to pass through it. However, it does not say anything about the direction of the induced voltage. Lenz's law is one of the basic laws of electromagnetic induction and helps to determine the direction of induced emf.

Lenz's law states that 'The direction of an induced emf is such that it will always oppose the change that is causing it.' This means that if the magnetic flux decreases, then the induced emf will oppose this decrease by generating an induced magnetic flux that adds to the original flux. Lenz's law, when combined with Faraday's, can be expressed as

$$e = -N\frac{\mathrm{d}\varphi}{\mathrm{d}t}. \tag{7.2}$$

The negative sign in equation (7.2) indicates that the direction of the induced emf is always opposite to its cause.

7.4 Mutual induction

The development of voltage in any electric circuit due to changing magnetic field is known as the induction effect. The changing magnetic field, if created by another electric circuit, leads to the theory of mutual inductance.

In order to understand the concept of mutual inductance, let us picture two coils A and B placed next to each other. An alternating voltage is connected across coil A, which results in alternating current and an alternating magnetic field around the coil. This alternating magnetic field when linked with coil B results in induced emf across coil B. The induced emf will cause a current to flow in coil B, which tries to maintain the existing magnetic field. The fact that the induced field always opposes the change is an example of Lenz's law.

The fact that a change in the current of one coil affects the current and voltage in the second coil is quantified in the property called mutual inductance.

The mutual inductance M can be defined as the proportionality between the emf generated in the secondary coil and the change in current in the primary coil that produced it.

If the current in coil A is I_1 and the flux through coil B of N_2 turns is ϕ_2, then

$$N_2\phi_2 \propto I_1$$

$$N_2\phi_2 = MI_1,$$

where M is a constant and is the mutual inductance of the two coils A and B.

For currents varying with time,

$$\frac{\mathrm{d}(N_2\phi_2)}{\mathrm{d}t} = \frac{\mathrm{d}(MI_1)}{\mathrm{d}t}. \quad (7.3)$$

As per Faraday's law the induced emf in coil B is given by

$$e = -\frac{\mathrm{d}(N_2\phi_2)}{\mathrm{d}t}. \quad (7.4)$$

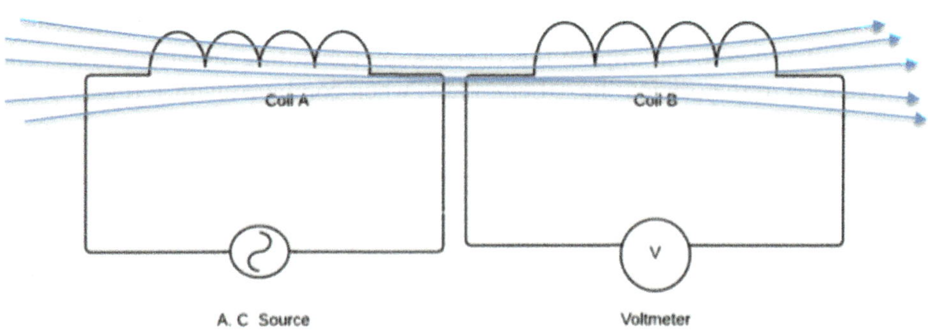

Figure 7.4. The mutual inductance effect.

On putting the value of $N_2\phi_2$ from equation (7.3) into equation (7.4), we obtain

$$e = -\frac{\mathrm{d}(MI_1)}{\mathrm{d}t}$$
$$e = -M\frac{\mathrm{d}I_1}{\mathrm{d}t}.$$

This shows that varying current in a coil can induce emf in a neighboring coil. The magnitude of the induced emf depends upon the rate of change of the current and the mutual inductance of the two coils.

The mutual inductance between two coils depends upon several factors:

(a) the number of turns in each coil
(b) the distance between two coils
(c) the orientation of each coil

7.5 Self-induction

In the previous section, we considered the flux in one coil due to the current in the other. It is also possible for emf to be induced in a single coil by changing the flux through it by means of varying the current in the same coil. This phenomenon is called self-induction. The concept of self-induction can be understood clearly by applying the concept of equilibrium given by the French chemist Henry Louis Le Châtelier. He established the 'equilibrium law' to predict the effect of a change on a chemical equilibrium. In fact, this principle is not only applicable to chemistry; it is a universal equilibrium law.

The generalized version of Le Châtelier's principle is stated as:

Any change in the existing state of affairs of a system prompts an opposing reaction in the responding system.

That is, if one tries to change the conditions of a system, the system will respond in a way so as to minimize the effect of change.

As per Lenz's law, 'The direction of an induced emf is such that it will always oppose the change that is causing it.' Thus Lenz's law is a phenomenon observed in an electric/magnetic system and is equivalent to the Le Châtelier principle.

In order to understand the concept of self-inductance, let us consider a coil of N turns connected across an AC source. The flux linkage through the coil is proportional to the current through the coil and is expressed as

$$\begin{aligned} N\phi &\propto I \\ N\phi &= LI, \end{aligned} \qquad (7.5)$$

where the constant of proportionality L is called the self-inductance of the coil, with units of Henry (H) after the American scientist Joseph Henry. When the current is varied, the flux linked with the coil also changes and an emf is induced in the coil. The induced emf opposes the cause, i.e. the change in flux, and is given by

$$e = -\frac{\mathrm{d}(N\phi)}{\mathrm{d}t}.$$

On putting in the value of $N\phi$ from equation (7.5), we obtain

$$e = -\frac{\mathrm{d}(LI)}{\mathrm{d}t}$$
$$e = -L\frac{\mathrm{d}I}{\mathrm{d}t}.$$

Thus, the self-induced emf always opposes any change of current in the coil.

Self-inductance is a property of the coil and can be estimated from its geometry. Let us calculate the self-inductance of a long coil of cross-sectional area A and length l, with $n = N/l$ turns per unit length carrying current I. The magnetic field due to current I flowing in the coil is $B = \mu_0 n I$. The total flux linked with the coil is

$$N\phi = nlBA.$$

On putting in the value of B, we obtain

$$N\phi = nl(\mu_0 nI)A$$
$$N\phi = n^2 l\mu_0 IA. \tag{7.6}$$

On comparing equations (7.5) and (7.6), the self-inductance is

$$L = n^2\mu_0 lA.$$

If we wind the coil on a magnetic material of relative permeability μ_r, the self-inductance of the coil, which also depends upon the permeability of the medium, is

$$L = n^2\mu_0\mu_r lA.$$

Physically, the self-inductance plays the role of inertia and is the electromagnetic analogue of mass in mechanics. So, work needs to be done against the back emf (e) in establishing the current. This work done is stored as magnetic potential energy. For the current I at an instant in a circuit, the rate of work done is

$$\frac{\mathrm{d}W}{\mathrm{d}t} = eI.$$

If we ignore the resistive losses and only consider inductive effect, then

$$\frac{\mathrm{d}W}{\mathrm{d}t} = LI\frac{\mathrm{d}I}{\mathrm{d}t}.$$

The total amount of work done in establishing the current I is

$$W = \int \mathrm{d}W = \int (LI)\mathrm{d}I.$$

Thus, the energy required to build up the current I is

$$W = \frac{1}{2}LI^2.$$

This expression is very similar to the kinetic energy of a particle of mass $m(= mv^2/2)$ and shows that L is analogous to mass, i.e. L is electrical inertia and opposes the growth and decay of current in the circuit.

7.6 The concept of an inductor

An inductor is a passive electrical element—a coil of wire wound on a solid iron, soft ferrite or hollow (free air) core—that increases the concentration of a magnetic field in a small physical space.

As discussed earlier, a current in an inductor (coil) produces a magnetic flux that is proportional to it. An inductor exhibits the property of inductance as it opposes the rate of change of the current flowing through it due to the build-up of energy in its magnetic field.

Let us consider a simple circuit with an AC source $V(t)$ and an inductor L with current $I(t)$ flowing through it. The current $I(t)$, which varies with time, induces a back emf $V_L(t)$, as given by equation (7.7), in the coil.

$$V_L(t) = -L\frac{\mathrm{d}I}{\mathrm{d}t}. \tag{7.7}$$

Thus the magnitude of V_L is proportional to the rate of change of the current and the polarity is opposite to that of the source voltage, as indicated in figure 7.5.

From equation (7.7), one can say that an inductor has an inductance of one Henry if it induces an emf of one volt when the current flowing through it changes at a rate of 1 A s^{-1}. However, if the current flowing through the inductor is constant,

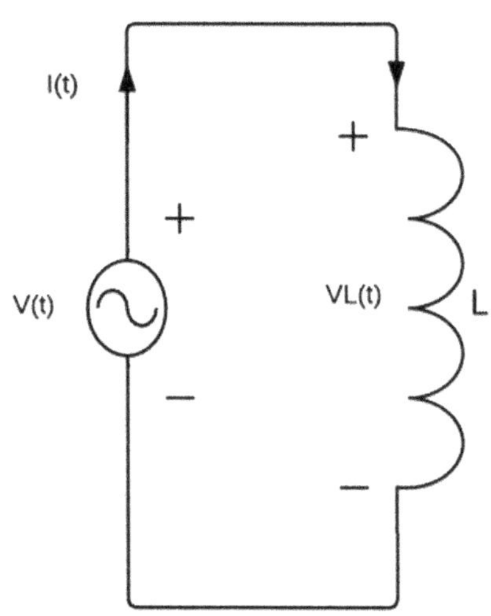

Figure 7.5. Inductor with an AC source.

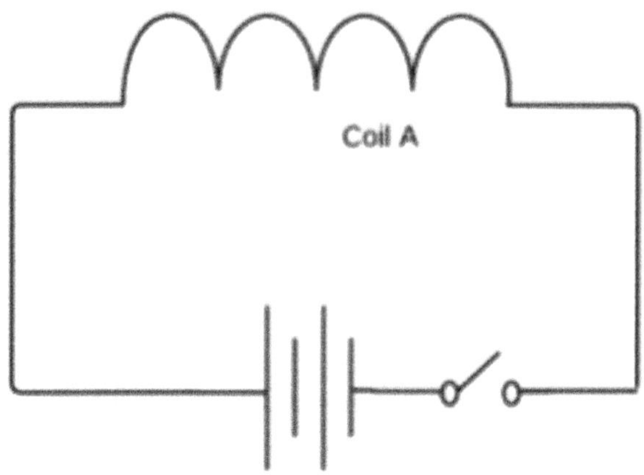

Figure 7.6. An inductor with a DC source and a switch S.

i.e. DC, the voltage induced in the inductor is zero. Zero induced voltage indicates that an inductor (ideal) acts as a short-circuit when subjected to DC current.

An inductor does not allow the current flowing through it to change instantaneously, since any instantaneous change in current will lead to an infinite rate of change of the current, $\mathrm{d}i/\mathrm{d}t = \infty$, making the induced emf infinite as well, and it is a known fact that infinite voltage does not exist. However, a sudden change in the inductor current results in a high voltage being induced across the inductor. This phenomenon can be explained with the help of the circuit diagram shown in figure 7.6.

When the switch S is open, no current flows through the coil A and thus the induced emf is also zero. On closing the switch S, a current flows through the circuit and slowly rises to the maximum value at a rate decided by the inductance L of the coil. This current results in an induced emf with the magnitude $L(\mathrm{d}i/\mathrm{d}t)$. This induced emf opposes the applied voltage until the current reaches its maximum value and a steady state condition is achieved. Only the resistance of the coil determines the current in the steady state condition as the reactance of the coil reduces to zero. Now, on opening the switch S, the current flowing through it falls to zero; the rate of the decrease is again decided by the inductance of the coil. This change in current induces voltage, which tries to maintain the current at the previous value. Thus with decreasing current the polarity of the induced voltage acts like a source, and with increasing current the voltage polarity acts as a load, as shown in figure 7.7.

7.7 Energy stored in an inductor

As discussed in the previous section, we know that an inductor opposes a change in current by inducing emf across it. The external supply works in order to maintain the flow of current against the induced emf and thus provides power for it. The instantaneous power provided by the supply to maintain the current I against the induced emf V_L is given as

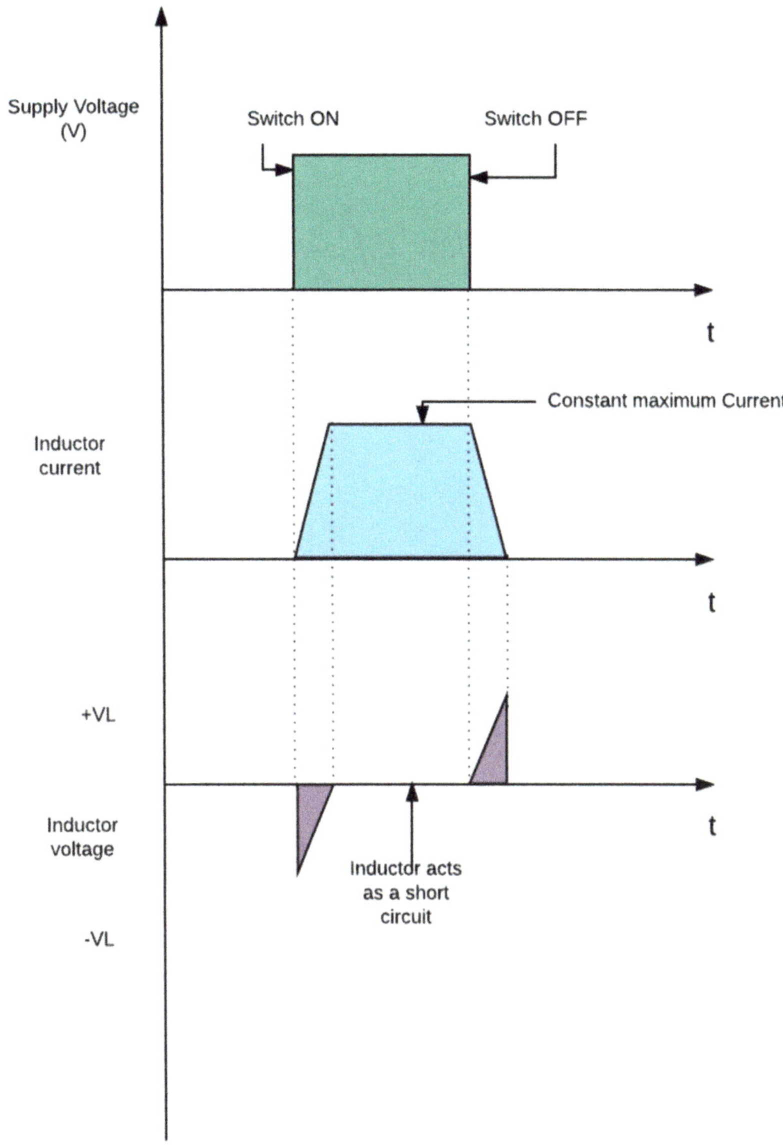

Figure 7.7. Current and voltage waveform across an inductor for the circuit shown in figure 7.6.

$$P = V_L I = \left(L\frac{\mathrm{d}I}{\mathrm{d}t}\right)I = \frac{1}{2}L\frac{\mathrm{d}I^2}{\mathrm{d}t}$$

$$P = \frac{\mathrm{d}}{\mathrm{d}t}\frac{1}{2}LI^2.$$

In the case of an ideal inductor, the resistance is zero and thus the power dissipated in a coil is zero. The power supplied by the source results in the storage of energy in the magnetic field. If the current in the circuit is increasing, i.e. $\mathrm{d}I/\mathrm{d}t > 0$, then $P > 0$.

This indicates that the work done by the external source is positive and results in the increase of the internal energy of the inductor. On the other hand, if the current in the circuit is decreasing, $\mathrm{d}I/\mathrm{d}t < 0$, then $P < 0$. This indicates that the external source is taking energy from the inductor and thus causing a reduction to the internal energy of the inductor. The total work W done by the external source in increasing the current from zero to I is

$$W = \int \mathrm{d}W = \int_I^0 \frac{1}{2}LI^2.$$

This is the energy stored within the magnetic field that surrounds the inductor by the current flowing through it.

Thus in an AC circuit, an inductor is constantly storing and providing energy in each cycle. In the case of a DC circuit, since $\mathrm{d}I/\mathrm{d}t = 0$, $P = 0$, resulting in no change in energy.

Conclusion

In this chapter, we have looked at Faraday's experiments and seen how the equation he formulated correctly predicts the results of the experiments. We have seen how to determine the direction of current by applying Lenz's law. We have also understood the concept of self- and mutual inductance and the various properties of a passive component called inductor.

Exercises

MCQs

1) A bar magnet is released from rest along the axis of an infinitely long vertical copper tube. After sufficient time the magnet will:
   a) Stop
   b) Move with almost constant velocity
   c) Accelerate downward continuously
   d) Oscillate
2) Consider the circuit shown in figure M7.1. A conducting loop ABCD is placed next to the circuit. When the switch S has been closed and then after some time opened again, the closed loop will show:

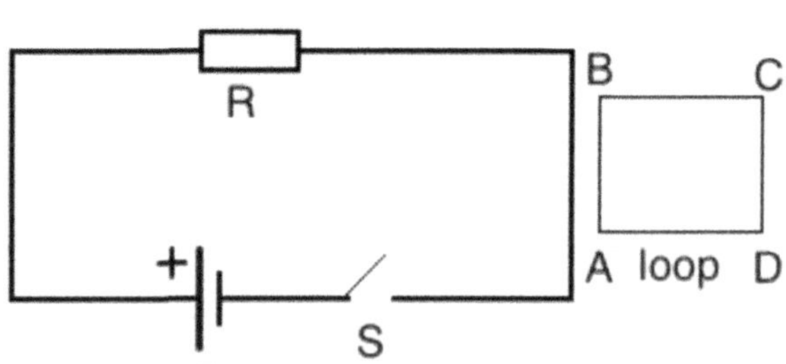

Figure M7.1.

   a) A clockwise current pulse
   b) An anti-clockwise current pulse
   c) A clockwise current pulse and an anti-clockwise current pulse
   d) An anti-clockwise current pulse and a clockwise current pulse
3) A conducting loop is placed in a magnetic field B. Which of the following processes does not induce an emf in the loop?
   a) The loop is translated from one point in space to another
   b) The loop is rotated about a diameter
   c) The loop is deformed
   d) The loop is stretched and its area increases
4) Moving a conductor in a constant magnetic field can induce an emf in the conductor. Is this statement true or false?
5) The magnetic field energy in an inductor changes from the maximum value to the minimum value in 5 ms. The frequency of the AC source is:
   a) 20 Hz
   b) 30 Hz
   c) 40 Hz
   d) 50 Hz

Problems

1) A rectangular loop ABCD is placed in a magnetic field B such that the side AB is in the field and the side CD has a resistor or resistance R. The field is parallel to the axis of the loop. A force F is applied to the loop and the loop is pulled out with a uniform speed v. Find the power dissipated in the resistor.
2) An average induced emf of 1 V appears in a coil when the current in it is changed from 1 A in one direction to 1 A in the opposite direction in 1 s. Find the self-inductance of the coil.
3) Calculate the energy stored in an inductor of inductance 100 mH when a current of 1 A is passed through it.
4) A conducting circular loop is placed in a uniform magnetic field $B = 0.01$ T with its plane perpendicular to the field. Let us suppose now that the radius of the loop starts increasing at a rate of 1 mm s^{-1}. Find the induced emf in the loop at an instant when the radius is 1 cm.
5) Consider a circuit in which an inductance L and a resistor R are connected to a battery of emf E. Find the maximum rate at which energy is stored in the magnetic field.

IOP Publishing

The Foundations of Electric Circuit Theory

N R Sree Harsha, Anupama Prakash and D P Kothari

Chapter 8

Maxwell's equations

8.1 The finite current-carrying wire

We have studied before that the Ampère's law and the Biot–Savart's law are well known tools to calculate the magnetic field created around current-carrying conductors. The former is used in high symmetry situations and the latter is used in a more general way. In this chapter, we shall understand why Ampère's law cannot be applied to a finite current-carrying wire while the Biot–Savart's law can be applied. In doing so, we shall also see how the charge conservation is implicitly taken into account in the Biot–Savart's law.

Consider a finite wire AB of length $2L$ carrying a current I, as shown in figure 8.1. We saw in a chapter 6 that Biot–Savart's law can be used to calculate the magnetic field created by the finite wire AB at a point P. We also stated that Ampère's law cannot be used in this situation to calculate the magnetic field because a finite wire is an unphysical situation. In this section, we shall see the results obtained by applying these two laws and try to add a term to Ampère's law to yield the correct result.

Using Biot–Savart's law

Biot–Savart's law is based on the principle that currents, which are the motion of charges, are the source of magnetic fields. The wire AB can be broken down into infinitesimal segments and the contribution from each segment should be superposed to calculate the magnetic field from the entire wire, as shown in figure 8.2. The result obtained was

$$\vec{B} = -\frac{\mu_0 I}{2\pi x}\frac{L}{\sqrt{L^2 + x^2}}\hat{k}. \tag{8.1}$$

Equation (8.1) is the magnetic field intensity at point P at a distance x from the midpoint of the wire AB. Let us now try to calculate the same problem with Ampère's law and see if this gives us the correct result.

doi:10.1088/978-0-7503-1266-0ch8

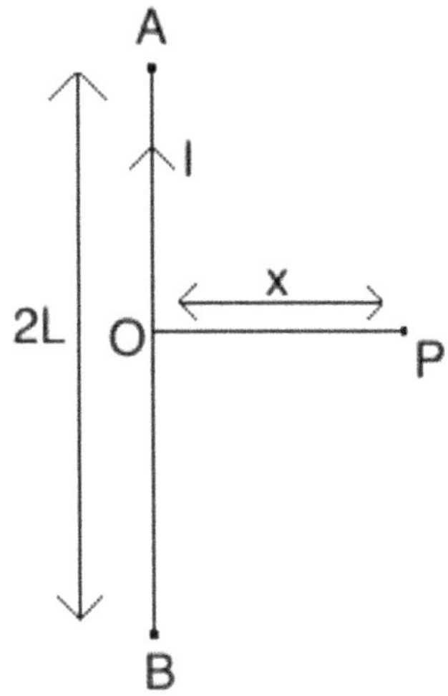

Figure 8.1. A finite wire *AB* of length $2L$ carrying a finite current *I*. The magnetic field at point *P* is calculated using Biot–Savart's law. Ampère's law cannot be used to calculate the magnetic field at *P* because a finite current-carrying wire is an unphysical situation.

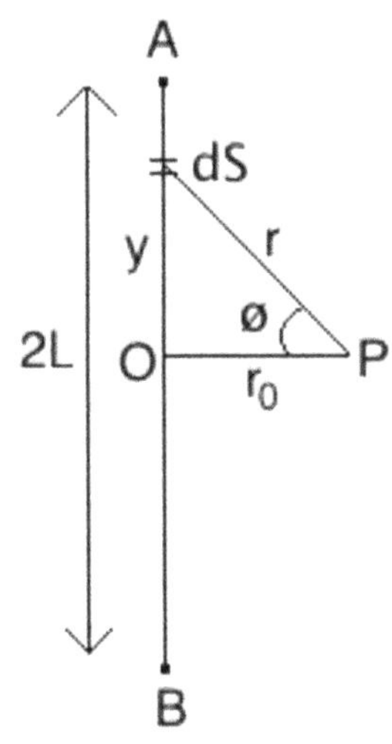

Figure 8.2. The finite current element is broken down into infinitesimal current elements, such as d**S**.

Using Ampère's circuital law

Consider an infinite long thin wire carrying a current *I*. Suppose the wire is oriented along the *y*-axis. We draw a circular loop around the wire that is centered at the wire and has a radius *x*. Then, Ampère's law states that the circulation of the magnetic vector **B** around this circle (or any closed contour for that matter) is equal to the product of μ_0 and the algebraic sum of currents enveloped by the circle:

$$\oint \vec{B} \cdot \mathrm{d}\vec{l} = \mu_0 \Sigma I.$$

For our case of an infinite wire, we have the magnitude of the magnetic vector **B** at any point in the circle as

$$B = \frac{\mu_0 I}{2\pi x}. \tag{8.2}$$

Now, Ampère's law gives the same result for a wire AB of finite length. We must expect from equation (8.1) that the magnitude of the magnetic field from the finite wire should depend on the length of the wire. But, as seen from equation (8.2), the magnetic field calculated from Ampère's law does not include the dimensions of the wire. Thus, equation (8.2) is not valid for calculating the magnitude of the magnetic field for a finite current-carrying wire. Hence, we must modify Ampère's law to obtain the result given in equation (8.1) for this particular situation.

8.2 Discharging a capacitor problem

Let us now study another example where Ampère's law gives us an incorrect result. Consider the circuit shown in figure 8.3. An ideal charged parallel place capacitor with capacitance C is being discharged through an external ohmic resistor R.

The energy is stored in the charged capacitor in the form of the electric field. When it is connected to an external resistor it is converted to heat in the resistor and dissipated into the environment. Let us take a closer look at how the capacitor in figure 8.3 discharges.

Let us draw a circular contour T, as shown in figure 8.4. The current-carrying wire passes through the center of the contour.

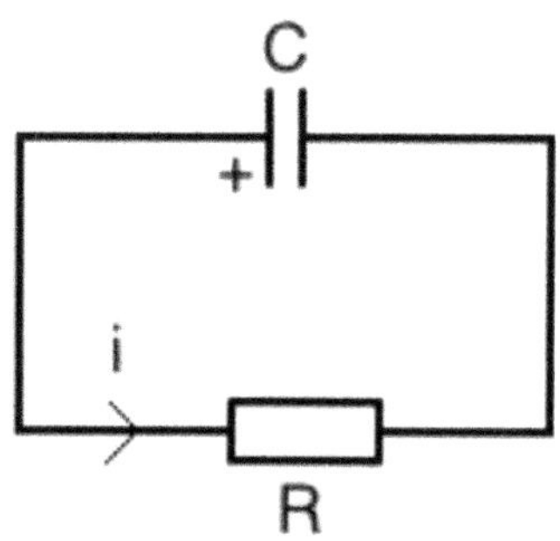

Figure 8.3. A capacitor discharging its charge through a resistor R.

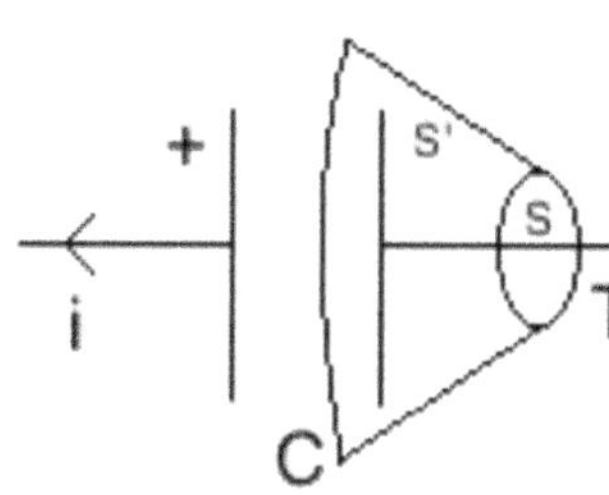

Figure 8.4. Surfaces S and S' are drawn attached to contour T.

Let us now attach two surfaces to the contour T. One of the surfaces, S, cuts the wire at its center. The other surface, S', does not cut through the wire. It passes through the negative plate. These two surfaces must yield equal results when Ampère's law is applied, because Ampère's law does not give special importance to any surface.

Firstly, using surface S for Ampère's law yields that the magnetic field B is created as

$$\int_T \vec{B} \cdot \mathrm{d}\vec{l} = \int_S \nabla \times \vec{B} \cdot \mathrm{d}\vec{a} = \mu_0 i.$$

Here, **dl** represents the line element in the contour T and **da** represents surface element in the surface S. Next, let us apply the same line of reasoning with surface S'. Clearly, no current cuts through surface S' and hence we get that

$$\int_T \vec{B} \cdot \mathrm{d}\vec{l} = \int_{S'} \nabla \times \vec{B} \cdot \mathrm{d}\vec{a} = 0.$$

This is an inconsistent result. It means that the magnitude of the magnetic field B depends on the surface chosen to calculate it. This also means that Ampère's law cannot be correct for a system in which charge density varies with time. Hence, Ampère's law must be extended to include time-varying current and charge densities. In the next section we shall see how by adding another term called 'displacement current' to Ampère's law we can obtain a more general equation in which charge densities can vary with time.

8.3 Concept of displacement current

Maxwell first introduced the term displacement current in 1862, purely on theoretical grounds. He does not tell us explicitly how he arrived at this term in his *Treatise*. It is often, however, incorrectly believed that he used the symmetry argument of electric and magnetic fields [1]. We see no such arguments in Maxwell's texts. Displacement current seems to have arisen while he was formulating the magnetic vorticity law. Maxwell used a mechanical model consisting of vortices and idle wheels to explain and derive his laws of electromagnetism. This is, however, very difficult and we shall now use the symmetry argument to discover the displacement current term.

The symmetry argument states that a time-varying electric field creates a time-varying magnetic field. We have seen this in Faraday's laws of induction. It should, therefore, be expected that a time-varying magnetic field will create a time-varying electric field.

Returning to the discharging of a capacitor problem, we see that surface S cuts a current i. Hence, we can write Ampère's law as

$$\int_T \vec{B} \cdot \mathrm{d}\vec{l} = \int_S \nabla \times \vec{B} \cdot \mathrm{d}\vec{a} = \mu_0 i = \mu_0 \int_S \vec{J} \cdot \mathrm{d}\vec{a}.$$

Here, **J** represents the current density flowing through the wire. Since surface S is common in the integral, we can write that

$$\nabla \times \vec{B} = \mu_0 \vec{J}. \tag{8.3}$$

This is called the differential form of Ampère's law. Consider now surface S'. It is clear that surface S' cuts a time-varying electric field. Hence, one more term should be added to equation (8.3) to reflect this. Let us write

$$\nabla \times \vec{B} = \mu_0 \vec{J} + X. \tag{8.4}$$

We have to now find out what X is. We have sufficient information to reveal X. We know that the continuity equation requires that

$$\nabla \cdot \vec{J} + \frac{\partial \rho}{\partial t} = 0.$$

And we also know that the divergence of a curl is zero. So, taking the divergence of equation (8.4), we obtain that

$$\nabla \cdot \left(\nabla \times \vec{B}\right) = \mu_0 \nabla \cdot \vec{J} + \nabla \cdot X = 0.$$

Comparing the two equations, we obtain that

$$\frac{1}{\mu_0} \nabla \cdot X = \frac{\partial \rho}{\partial t}.$$

From the differential form of Coulomb's law, we know that $\nabla \cdot E = \frac{\rho}{\varepsilon_0}$. Hence, we obtain that

$$\frac{1}{\mu_0 \varepsilon_0} \nabla \cdot X = \nabla \cdot \frac{\partial \vec{E}}{\partial t}.$$

Upon rearranging, we see that

$$\nabla \cdot X = \nabla \cdot \left(\mu_0 \varepsilon_0 \frac{\partial \vec{E}}{\partial t}\right).$$

Thus, we have derived the quantity X. It is equal to $\mu_0 \varepsilon_0 \frac{\partial \vec{E}}{\partial t}$. Now, the full differential form of Ampère's law can be written as

$$\begin{aligned} \nabla \times \vec{B} &= \mu_0 \vec{J} + \varepsilon_0 \mu_0 \frac{\partial \vec{E}}{\partial t} \\ \Rightarrow \nabla \times \vec{B} &= \mu_0 \left(\vec{J} + \varepsilon_0 \frac{\partial \vec{E}}{\partial t}\right) = \mu_0 \left(\vec{J} + \vec{J}_d\right). \end{aligned} \tag{8.5}$$

Equation (8.5) is called the Maxwell–Ampère law. Maxwell called the term $\mathbf{J}_d$ the displacement current. The magnitude and the direction of this is given by

$$\vec{J}_d = \varepsilon_0 \frac{\partial \vec{E}}{\partial t}.$$

8.3.1 Solution to the discharging capacitor problem

Let us now apply this term to our surface S' attached to contour T and see if it gives us the correct result. We have to prove that the current passing through surface S' is equal to the current passing through S. Let us now consider that the two surfaces S and S' form a closed volume. The divergence of the electric field $\mathbf{E}$ from this closed volume is given by

$$\nabla \cdot \vec{E} = \frac{\rho}{\varepsilon_0}.$$

Here, ρ represents the time-dependent charge in the volume enclosed by surfaces S and S'. Taking the time derivative, we have that

$$\nabla \cdot \frac{\mathrm{d}\vec{E}}{\mathrm{d}t} = \frac{1}{\varepsilon_0} \frac{\mathrm{d}\rho}{\mathrm{d}t}. \tag{8.6}$$

We also have from the continuity equation that

$$\nabla \cdot \vec{J} + \frac{\mathrm{d}\rho}{\mathrm{d}t} = 0.$$

Here $\mathbf{J}$ represents the current density that is carried away by the conducting wires cutting through surface S. Using this continuity equation in equation (8.6), we obtain that

$$\nabla \cdot \left(\varepsilon_0 \frac{\partial \vec{E}}{\partial t} \right) = -\nabla \cdot \vec{J}. \tag{8.7}$$

Since the differential form of equation (8.7) represents the relationship between the values of the local electric fields and the current density, we can write

$$\varepsilon_0 \frac{\partial \vec{E}}{\partial t} = -\vec{J} = \vec{J}_d.$$

Hence, we have proved that the magnitude of the current density is equal to the magnitude of the displacement current density. We can understand the sign of the displacement current being opposite in sign to the actual current in the wire by noting that the direction of normal to surface S' is opposite to the direction of normal to surface S in the closed volume enclosed by S and S'. Now, if we direct the normal for surface S' to the same side as for surface S, we obtain that

$$\vec{J} = \vec{J}_d.$$

Hence, we have proved that our capacitor problem can be completely solved by introducing a term called displacement current to Ampère's law.

8.3.2 Solution to the finite current-carrying wire problem

T Charitat and F Graner first proposed this solution in their brilliant paper [2]. Consider the finite wire AB, as shown previously in figure 8.1. One simple explanation is that the finite wire does not represent a physically correct situation because current always needs a closed path to flow. So, Ampère's law gives us an incorrect result because of this misapplication. However, for an infinite wire, the wire can be thought of as closed at infinity and hence Ampère's law in an infinite current-carrying wire gives us the correct result. Let us now use Maxwell–Ampère's law to derive the magnetic field produced by the finite wire AB.

Consider the wire shown in figure 8.5. We have a finite wire carrying a current I and we also have sources and sinks of positive charge at B and A, respectively. Consider a point P at a distance x from the mid-point of the wire. The source is represented by $q(t)$ and the sink by $-q(t)$. Consider a contour T and let us stretch a surface S on the contour.

We have the differential form of the Maxwell–Ampère law as

$$\nabla \times \vec{B} = \mu_0 \vec{J} + \mu_0 \varepsilon_0 \frac{\partial \vec{E}}{\partial t}. \tag{8.8}$$

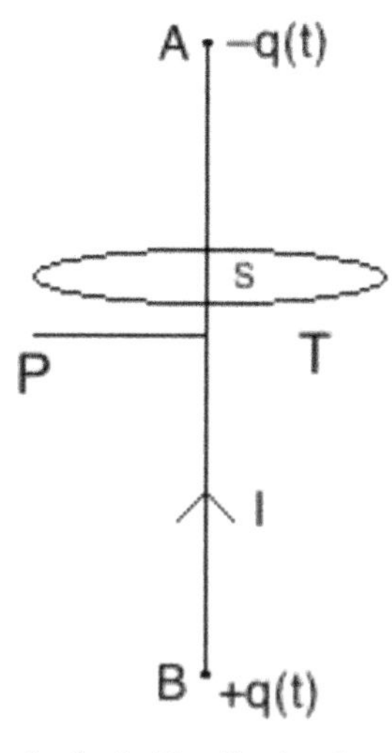

Figure 8.5. The finite wire element is made physical by allowing the sources and sinks of positive charges at B and A, respectively.

Next, integrating equation (8.8) with small elements, dS of the surface **S** as the variables of integration, we obtain that

$$\int_S \nabla \times \vec{B} \cdot \mathrm{d}\vec{S} = \mu_0 \int_S \vec{J} \cdot \mathrm{d}\vec{S} + \varepsilon_0 \mu_0 \frac{\partial}{\partial t}\left(\int_S \vec{E} \cdot \mathrm{d}\vec{S} \right).$$

Using Stoke's theorem we obtain that

$$\oint_T \vec{B} \cdot \mathrm{d}\vec{l} = \mu_0 I + \mu_0 \varepsilon_0 \frac{\partial \Phi_E}{\partial t}. \tag{8.9}$$

Here, Φ_E represents the electric flux through the surface. It can be calculated as

$$\Phi_E = \int_S \vec{E} \cdot \mathrm{d}\vec{S}.$$

We need to calculate this electric flux that cuts surface S. Let us, to simplify the problem, center the contour T and the surface S at the middle point of the finite wire AB. The contour is now a circle with a radius x. We will consider this point to be the origin and point A then has its coordinate y as $y = L$ and point B has the coordinate $y = -L$.

We have an electric dipole whose electric field is to be calculated at its center. The y-component of the electric field is given by

$$E_y = \frac{qL}{2\pi\varepsilon_0 \left(x^2 + L^2\right)^{3/2}}.$$

The electric flux bounded by the flat area centered at the origin and at a distance of x is given by

$$\Phi_E = \int_0^X E_y \, 2\pi r \, \mathrm{d}r.$$

Upon integration, we have that

$$\Phi_E = \frac{q}{\varepsilon_0}\left[1 - \frac{L}{\left(L^2 + x^2\right)^{0.5}}\right].$$

Now, we have that

$$\frac{\mathrm{d}\Phi_E}{\mathrm{d}t} = \frac{-I}{\varepsilon_0}\left[1 - \frac{L}{\left(L^2 + x^2\right)^{0.5}}\right].$$

Substituting these results into equation (8.9), we obtain that

$$\oint_T \vec{B} \cdot \mathrm{d}\vec{l} = \mu_0 I + \mu_0 \varepsilon_0 \left[-\frac{I}{\varepsilon_0}\left(1 - \frac{L}{\sqrt{L^2 + x^2}}\right)\right].$$

Hence, we have

$$\oint_T \vec{B} \cdot \mathrm{d}\vec{l} = 2\pi x \, |\vec{B}| = \frac{\mu_0 I L}{\sqrt{L^2 + x^2}}$$

$$\Rightarrow |\vec{B}| = \frac{\mu_0 I L}{2\pi x \sqrt{L^2 + x^2}}.$$

Compare this equation with equation (8.1), which was obtained using Biot–Savart's law. Hence, we have demonstrated the validity of the Maxwell–Ampère law for the case of a finite current-carrying wire.

Example problem

Question

Consider a radioactive source S that is emitting positive particles at a constant rate in all directions. Consider a Gaussian surface G surrounding the source as shown in figure 8.6.

Now, consider the magnetic field **B** created by the particles ejected by the source. Gauss's law for magnetism implies that the divergence of the magnetic field **B** is always zero. Hence, at any point on the surface, if there is a radial outward magnetic field, somewhere else there has to be a radial inward magnetic field of the same magnitude to cancel the outward field in the Gaussian summation. However, the entire Gaussian surface is a sphere and has a spherical symmetry. Hence, we conclude that there will not be any radial component of the magnetic field on the Gaussian surface. Next, consider two Amperian loops A_1 and A_2 that touch at point P, as shown in figure 8.6. Calculate the tangential component of the magnetic field using these Amperian loops.

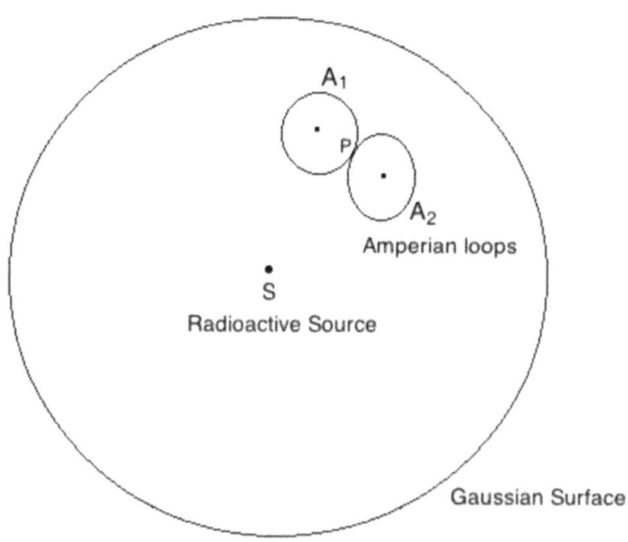

Figure. 8.6. A diagram showing Gaussian surface around a radioactive source.

Solution

It was already demonstrated that the radial component of the magnetic field should be zero. Let the radius of the Amperian loop be r. By applying Ampère's law to the circular loop A_1, we obtain that

$$\oint_{A_1} \vec{B} \cdot \mathrm{d}\vec{l} = 2\pi r B_{\|} = \mu_0 I.$$

Now, as seen from the outside of the Gaussian surface, the direction of $B_{\|}$ is counter-clockwise. Next, consider the second Amperian loop A_2 with same radius r. Since both loops have the same radius, equal currents I pass through them. Thus, we also have that

$$\oint_{A_2} \vec{B} \cdot \mathrm{d}\vec{l} = 2\pi r B_{\|} = \mu_0 I.$$

The magnetic field produced by this loop is also in the counter-clockwise direction, as seen from the outside of the Gaussian surface. Hence, at point P the two equal and opposite tangential magnetic fields cancel and we have that the magnetic field is zero at point P. And if we take the radius of loop A_1 to be larger than the radius of loop A_2, we obtain the tangential component of the magnetic field at point P in one direction, while if we take the radius of loop A_1 to be smaller than that of A_2, it is in the opposite direction. Hence, we arrive at the strange and nonsensical conclusion that the magnetic field depends on the radius of the Amperian loop. Let us see how the introduction of displacement current solves this problem.

The time-dependent electric flux through surface G, as we have already seen, is given by

$$\Phi_E = \frac{q(t)}{\varepsilon_0}.$$

Differentiating the above equation with respect to time, we get that

$$\frac{\mathrm{d}\Phi_E}{\mathrm{d}t} = \frac{1}{\varepsilon_0}\frac{\mathrm{d}q(t)}{\mathrm{d}t}. \tag{8.10}$$

The electric flux though the Amperian loop A_1 is a proportional fraction of this total electric flux. Hence, we have

$$\left.\frac{\mathrm{d}\Phi_E}{\mathrm{d}t}\right|_{A_1} = \frac{\mathrm{d}\Phi_E}{\mathrm{d}t}\frac{A}{4\pi r^2}.$$

Now, $\mathrm{d}q/\mathrm{d}t$ implies that the charge inside the Gaussian sphere is decreasing the continuity equation, which implies that this corresponds to the current I_t that flows out of the sphere. Hence, we have that

$$I_t = -\frac{\mathrm{d}q(t)}{\mathrm{d}t}.$$

Next, the current that flows through the Amperian loop A_1 with area A is I given by

$$I = I_t \frac{A}{4\pi r^2}.$$

With these, we can write equation (8.10) as

$$\frac{\mathrm{d}\Phi_E}{\mathrm{d}t} = -\frac{I_t}{\varepsilon_0} = -\frac{4\pi r^2 I}{A\varepsilon_0}.$$

We have the Maxwell–Ampère law that states that for loop A_1,

$$\oint_{A_1} \vec{B} \cdot \mathrm{d}\vec{l} = \mu_0 I + \mu_0 \varepsilon_0 \left.\frac{\mathrm{d}\Phi_E}{\mathrm{d}t}\right|_{A_1}.$$

Upon substitution of the appropriate values, we obtain that

$$\oint_{A_1} \vec{B} \cdot \mathrm{d}\vec{l} = B_{\|} 2\pi r = \mu_0 (I - I) = 0.$$

This implies that $B_{\|} = 0$, as required by the problem. Hence the inconsistency is removed by using the Maxwell–Ampère law, because the situation involves time-dependent sources. This brilliant problem was first proposed by Gary Reich [3].

8.4 Maxwell's equations

With the introduction of the displacement current, Maxwell completed the macroscopic theory of electricity and magnetism, unifying them into the single framework now known as 'classical electromagnetism'. We have only presented the equations that represent the individual properties of electric and magnetic fields. In this section, we shall present the complete set of Maxwell's equations and explain their properties.

Maxwell, inspired by the field concept proposed by Faraday, originally used the molecular vortex model as a mechanical model to understand electromagnetism. William Thomson (Lord Kelvin) first introduced this molecular vortex model in 1856. Maxwell employed it to explain the theory of electricity and magnetism from the field-primacy standpoint. Maxwell originally had a set of 20 equations to explain his theory. Later, Oliver Heaviside mastered this Faraday–Maxwell approach and reduced it to the set of four equations now called Maxwell's equations.

Let us consider a closed contour T as a boundary line of a closed volume V. The volume V has S as its surface. Also, let $\mathbf{dS}$ represent the infinitesimal vector area of surface S and $\mathrm{d}l$ represent the infinitesimal vector element of contour T. We have used two different meanings of S to represent the closed surfaces, using a circle in the integral symbol as appropriate. Then Maxwell's four equations in integral form are:

1. $\oint_S \vec{E} \cdot \mathrm{d}\vec{S} = \frac{1}{\varepsilon_0} \oint_V \rho \, \mathrm{d}V$

2. $\oint_S \vec{B} \cdot d\vec{S} = 0$
3. $\oint_T \vec{E} \cdot d\vec{l} = -\frac{d}{dt}\int_S \vec{B} \cdot d\vec{S}$
4. $\oint_T \vec{B} \cdot d\vec{l} = \mu_0 \int_S \vec{J} \cdot d\vec{S} + \mu_0\varepsilon_0 \frac{d}{dt}\int_S \vec{E} \cdot d\vec{S}$

Here, **J** represents the conduction current density and ρ represents the charge density. These equations represent concise knowledge about every experiment performed in electromagnetism. Let us now explain what each equation means.

1. The flux of the electric field **E** through a closed surface S is equal to the ratio of the total charge enclosed by the closed surface containing volume V to the ε_0. This is called Gauss's theorem.
2. The flux of the magnetic field **B** through a closed surface S is always zero. In other words, magnetic monopoles do not exist. This is called Gauss's law for magnetism.
3. The circulation of vector **E** along the closed contour T is equal to the negative of the time derivative of the magnetic flux passing through the open surface **S**. This is called Faraday's law.
4. The circulation of vector **B** along the closed contour T is equal to the μ_0 times the flux of the total (displacement and conduction) current through the open surface **S**. This is called the Maxwell–Ampère law.

It follows from these equations that electric and magnetic fields cannot be treated as independent quantities: a change in time in one of these fields leads to the emergence of the other. It should also be noted that these equations obtained by our line of reasoning should not be considered to be proof of the equations. Maxwell's equations are another way of writing the axioms of classical electromagnetism. To understand what we mean, consider Maxwell's second equation, which states that the flux of magnetic field through any closed surface S is always zero. This is another way of saying that there are no magnetic monopoles. The absence of magnetic monopoles is an observational fact that cannot be derived *a priori*.

It is also instructional to write Maxwell's equations in differential form. We have:

1. $\nabla \cdot \vec{E} = \frac{\rho}{\varepsilon_0}$
2. $\nabla \cdot \vec{B} = 0$
3. $\nabla \times \vec{E} = -\frac{\partial \vec{B}}{\partial t}$
4. $\nabla \times \vec{B} = \mu_0 \vec{J} + \mu_0\varepsilon_0 \frac{\partial \vec{E}}{\partial t}$

These Maxwell equations in differential form present the local relationship of electric and magnetic fields. Thus, we see here that the electric and magnetic fields are dependent and, with a little more calculation, we will show later that they exhibit a wave-like phenomenon that propagates at the speed of light c.

8.5 Helmholtz's theorem

We have yet to prove that the four Maxwell equations completely represent the electric and magnetic fields. In other words, we have to prove that if we know the divergence and curl of a vector field, we know everything there is to know about the field. This indeed is proved to be the case using Helmholtz's theorem, named after the German physicist and mathematician Hermann Ludwig Ferdinand von Helmholtz.

Suppose we have a scalar field ϕ and a field equation that tells us that

$$\nabla\phi = \vec{\psi}(\vec{r}). \tag{8.11}$$

Here, $\vec{\psi}(\vec{r})$ is a vector field that is a function of position in space. We have $\nabla \times \vec{\psi}(\vec{r}) = 0$ for consistency, since the curl of a gradient is always zero. Equation (8.11) completely specifies the scalar field ϕ once its value at a point P is given.

This is shown for an arbitrary point X by

$$\phi(X) = \phi(P) + \int_P^X \nabla\phi \cdot \mathrm{d}\vec{l} = \phi(P) + \int_P^X \vec{\psi}(\vec{r}) \cdot \mathrm{d}\vec{l}.$$

We have also that the curl of $\vec{\psi}(\vec{r})$ is equal to zero. This implies that $\vec{\psi}(\vec{r})$ is a conservative field and shows that the above equation gives a unique value for the scalar field $\phi(X)$ at X. Hence we have demonstrated that the gradient of a scalar field is sufficient to determine everything about that scalar field.

Let us now consider a vector field **F**. Suppose we also have that

$$\nabla \cdot \vec{F} = C$$
$$\nabla \times \vec{F} = \mathbf{D}.$$

We must also have for mathematical consistency that

$$\nabla \cdot \left(\nabla \times \vec{F}\right) = \nabla \cdot \vec{D} = 0.$$

Let us write the vector **F** as the sum of a conservative field and a solenoid field:

$$\vec{F} = -\nabla U + \nabla \times \vec{W}.$$

Taking the divergence and curl of the above equation, we have that

$$\nabla \cdot \vec{F} = -\nabla^2 U$$
$$\nabla \times \vec{F} = \nabla \times \left(\nabla \times \vec{W}\right) = \nabla\left(\nabla \cdot \vec{W}\right) + \nabla^2 \vec{W}.$$

In order to simplify things, let us consider the vector field **F** to be the electric field. We then have that

$$C = \frac{\rho}{\varepsilon_0}.$$

And

$$\vec{D} = -\frac{\partial \vec{B}}{\partial t}.$$

But, as we have seen in the Maxwell equations, in general the curl of electric fields is not equal to zero. It is in fact given by

$$\nabla \times \vec{E} = -\frac{\partial \vec{B}}{\partial t}. \tag{8.12}$$

Thus, for a time-independent case, we have that

$$\nabla \times \vec{E} = 0. \tag{8.13}$$

Equation (8.13) entails the existence of a potential function written as $\vec{E} = -\nabla\phi$.

But, we know that the divergence of the magnetic field is always zero, as shown by

$$\nabla \cdot \vec{B} = 0.$$

Hence, we can write the magnetic field **B** as a curl of a vector **A** because the divergence of a curl is always zero. This new vector **A** is called the magnetic vector potential.

$$\vec{B} = \nabla \times \vec{A}.$$

Hence, we can write equation (8.12) as

$$\nabla \times \vec{E} = -\frac{\partial\left(\nabla \times \vec{A}\right)}{\partial t},$$

which can be written as

$$\nabla \times \left(\vec{E} + \frac{\partial \vec{A}}{\partial t}\right) = 0.$$

Hence, even in a general time-varying case, we have effectively reduced equation (8.12) to an equation similar to (8.13). We can now write

$$\vec{E} + \frac{\partial \vec{A}}{\partial t} = -\nabla\phi.$$

Note that we are free to choose the divergence of the magnetic vector potential **A**. We employ the Coulomb gauge that states that the divergence of the magnetic vector potential is zero. That is,

$$\nabla \cdot \vec{A} = 0$$

$$\Rightarrow \nabla \cdot \frac{\partial \vec{A}}{\partial t} = 0.$$

Hence, we have reduced the electric field as the sum of a conservative field and a solenoid field. Similarly, for a general function **F**, it can be proved that **F** can be written as the sum of the solenoid and conservative fields and that $\nabla \cdot \vec{W} = 0$. Hence, we have

$$\nabla \cdot \vec{F} = -\nabla^2 U$$
$$\nabla \times \vec{F} = \nabla \times \left(\nabla \times \vec{W}\right) = \nabla^2 \vec{W}.$$

Now, the equation of the form $\nabla^2 g = 0$ is called the Laplace equation and it has unique solutions. Hence, given the unique solutions to the divergence and curl of a vector function, we can reconstruct the vector function without any ambiguity. This is called Helmholtz's theorem.

8.6 The choice of gauge

We have seen that electric and magnetic fields can be written in scalar and vector potentials as

$$\vec{E} = -\nabla\phi - \frac{\partial \vec{A}}{\partial t} \tag{8.14}$$

$$\vec{B} = \nabla \times \vec{A}. \tag{8.15}$$

However, only the curl of the magnetic vector potential **A** is defined as the magnetic field. In order to have a complete description of vector **A**, we also need to know the divergence of **A** according to the Helmholtz theorem. As we have already pointed out, we are free to choose the divergence of **A**. We are also free to choose the ground point for the electric potential because only the gradient of the potential enters into the equation for the electric field. Hence, we can add a constant term to the potential and it will not make any difference. In the same way, we can choose different values for the divergence of the magnetic vector potential and it will not make any difference. However, some properties are manifestly easily visible if we prefer to employ a particular gauge. Let us now consider two widely used gauges, namely the Coulomb and Lorentz gauges.

Coulomb gauge
If we set the divergence of the magnetic vector potential equal to zero, it is called the Coulomb gauge. Taking the divergence and curl of equations (8.14) and (8.15), respectively, we have that

$$\left.\begin{aligned} -\nabla^2\phi - \frac{\partial}{\partial t}\left(\nabla \cdot \vec{A}\right) &= \frac{\rho}{\varepsilon_0} \\ -\nabla^2\vec{A} + \nabla\left(\nabla \cdot \vec{A}\right) &= \mu_0\left[\vec{J} + \varepsilon_0\frac{\partial \vec{E}}{\partial t}\right] \end{aligned}\right\}. \tag{8.16}$$

Now, for the Coulomb gauge, we set $\nabla \cdot \vec{A} = 0$. Hence, equations (8.16) reduce to

$$\left.\begin{aligned} \nabla^2\phi &= -\frac{\rho}{\varepsilon_0} \\ \nabla^2\vec{A} &= -\mu_0\left[\vec{J} + \varepsilon_0\frac{\partial\vec{E}}{\partial t}\right] \end{aligned}\right\}. \tag{8.17}$$

The equation for the electric potential is exactly same as the steady state condition. When the sources are at the position vector represented by $\mathbf{r}'$, the potential at a point $\mathbf{r}$ is given by

$$\phi(\vec{r}, t) = \frac{1}{4\pi\varepsilon_0}\int \frac{\rho(\vec{r}\,', t)}{|\vec{r} - \vec{r}\,'|} \mathrm{d}^3\vec{r}\,'. \tag{8.18}$$

Note how the same time t comes in both the left- and right-hand sides of equation (8.18). This implies the action-at-a-distance formalism. It is a typical action-at-a-distance law for potential. We have seen in previous chapters that such a formalism is not consistent with special relativity. So, does this mean that the Coulomb gauge is wrong? No. The answer lies in the fact that scalar potentials cannot be measured directly. Only fields can be measured directly, and the potential is to be inferred from the fields. The electric field in the time-dependent case is shown by equation (8.14). It has two parts: the scalar potential part and the vector potential part. Even though the scalar potential changes instantaneously, as shown by equation (8.18), it is compensated by the equal negative change in the vector potential part so that the electric field remains constant. This state of affairs persists until sufficient time has elapsed for the signal to travel from the source to the point in question. In fact, in analogy with equation (8.18), the equation for the vector potential is given by solving the second equation in (8.17) as:

$$\vec{A}(\vec{r}, t) = \frac{\mu_0}{4\pi}\int \frac{\vec{J}(\vec{r}\,', t) + \vec{J}_d(\vec{r}\,', t)}{|\vec{r} - \vec{r}\,'|} \mathrm{d}^3\vec{r}\,'.$$

Hence, in this way, special relativity is not violated in the Coulomb gauge.

Lorentz gauge

The Lorentz gauge is when we set the divergence of $\mathbf{A}$ as follows:

$$\nabla \cdot \vec{A} = -\varepsilon_0\mu_0\frac{\partial\phi}{\partial t}.$$

Upon substitution into equation (8.16), we get that

$$\left.\begin{aligned} \varepsilon_0\mu_0\frac{\partial^2\phi}{\partial t^2} - \nabla^2\phi &= \frac{\rho}{\varepsilon_0} \\ -\nabla^2\vec{A} + \varepsilon_0\mu_0\frac{\partial^2\vec{A}}{\partial t^2} &= \mu_0\vec{J} \end{aligned}\right\}. \tag{8.19}$$

These equations (8.19) represent three-dimensional wave equations in which information propagates at the speed of light. Hence, the Lorentz gauge is the optimum gauge for time-dependent situations because it manifestly shows that information propagates at finite speeds c.

8.7 Retarded potentials and fields

It was previously stated that the potential function in Coulomb's gauge at a point r due to the charge density ρ present at r_0 is given by equation (8.18) as

$$\phi(\vec{r}, t) = \frac{1}{4\pi\varepsilon_0}\int \frac{\rho\left(\vec{r}', t\right)}{\left|\vec{r} - \vec{r}'\right|}\mathrm{d}^3\vec{r}'.$$

However, if time-dependent fields are also involved, then we need to take into account the limit for the speed at which information can travel in the universe, which is the speed of light c. Hence, equation (8.18) can be manifestly written as

$$\phi(\vec{r}, t) = \frac{1}{4\pi\varepsilon_0}\int \frac{\rho\left(\vec{r}', t - \frac{|\vec{r} - \vec{r}'|}{c}\right)}{\left|\vec{r} - \vec{r}'\right|}\mathrm{d}^3\vec{r}'. \tag{8.20}$$

Here, the quantity $t - \frac{|\vec{r} - \vec{r}'|}{c}$ is called the retarded time. Hence, only after a certain amount of time given by $\frac{|\vec{r} - \vec{r}'|}{c}$ is the information about the scalar potential function updated. This is the inherent restriction imposed by the laws of special relativity. Similarly, we also have for the vector potential

$$\vec{A}(\vec{r}, t) = \frac{\mu_0}{4\pi}\int \frac{\vec{J}\left(\vec{r}', t - \frac{\left|\vec{r} - \vec{r}'\right|}{c}\right)}{\left|\vec{r} - \vec{r}'\right|}\mathrm{d}^3\vec{r}'. \tag{8.21}$$

Equations (8.20) and (8.21) are the solutions of equation (8.19). Hence, we can see that the Lorentz gauge manifestly shows the speed limit of the universe imposed by special relativity.

8.8 Properties of Maxwell's equations

Linearity

It is immediately obvious from the differential form of Maxwell's equations that they only contain electric and magnetic fields as first-order time derivatives with the first power of sources ρ and **J**. This implies that the principle of superposition can be applied directly to these electric fields created by the sources.

Maxwell's equations include the continuity equation
This claim may not be very obvious at first glance. In order to prove it, let us consider the differential form of Maxwell's equations. Take the divergence of Maxwell's fourth equation to obtain

$$\nabla \cdot \left(\nabla \times \vec{B}\right) = \mu_0 \nabla \cdot \vec{J} + \mu_0 \varepsilon_0 \frac{\partial \nabla \cdot \vec{E}}{\partial t} = 0.$$

Next, using Maxwell's equation (1), we obtain that

$$\nabla \cdot \vec{J} + \varepsilon_0 \frac{\partial(\rho/\varepsilon_0)}{\partial t} = 0.$$

Upon simplification, we obtain the familiar continuity equation as

$$\nabla \cdot \vec{J} + \frac{\partial \rho}{\partial t} = 0.$$

The magnetic vector is a pseudo-vector
In this section, we shall see why the magnetic vector **B** is a pseudo-vector. We know that Maxwell's equations can be written in terms of the scalar and vector potentials as

$$\begin{aligned} \vec{B} &= \nabla \times \vec{A} \\ \vec{E} &= -\nabla \phi + \frac{\partial \vec{A}}{\partial t}. \end{aligned}$$

Defining the electric and magnetic fields this way guarantees that magnetic monopoles do not exist and that Faraday's law is automatically satisfied. They are as shown:

$$\begin{aligned} \nabla \times \vec{E} &= -\frac{\partial\left(\nabla \times \vec{A}\right)}{\partial t} = -\frac{\partial \vec{B}}{\partial t} \\ \nabla \cdot \vec{B} &= \nabla \cdot \left(\nabla \times \vec{A}\right) = 0. \end{aligned}$$

The scalar is a proper scalar and the vector potential is a proper vector. There is a rule that states that 'the curl of a proper vector is a pseudo-vector'. Hence, we have that **E** is a proper vector, but **B** is a pseudo-vector.

Time-reversal symmetry and the magnetic field
Let us demonstrate one more interesting property of Maxwell's equations. Consider the differential form of Maxwell's equations. If the direction of time is reversed, meaning that $t \to (-t)$, Maxwell's equations become

$$\nabla \cdot \vec{E} = \frac{\rho}{\varepsilon_0}$$
$$\nabla \cdot \vec{B} = 0$$
$$\nabla \times \vec{E} = +\frac{\partial \vec{B}}{\partial t}$$
$$\nabla \times \vec{B} = -\mu_0 \vec{J} - \varepsilon_0 \mu_0 \frac{\partial \vec{E}}{\partial t}.$$

The current density changes its sign because it is current per unit area, and we have current I defined as

$$I = \frac{\mathrm{d}Q}{\mathrm{d}t}.$$

Hence, we have that as $t \to (-t)$, $\vec{J} \to (-\vec{J})$. But, looking at Maxwell's equations, it is clear that as $\vec{B} \to (-\vec{B})$, the equations become:

$$\nabla \cdot \vec{E} = \frac{\rho}{\varepsilon_0}$$
$$\nabla \cdot \left(-\vec{B}\right) = 0$$
$$\nabla \times \vec{E} = -\frac{\partial\left(-\vec{B}\right)}{\partial t}$$
$$\nabla \times \left(-\vec{B}\right) = \mu_0 \vec{J} + \varepsilon_0 \mu_0 \frac{\partial \vec{E}}{\partial t}.$$

Now, these are just Maxwell's equations with the magnetic field being reversed in sign. Hence, we conclude that the laws of classical electromagnetism remain unchanged if everywhere time t is replaced by $-t$ and simultaneously **B** is replaced by **−B**. This property, although trivial, is useful in quantum electrodynamics and for proving certain theorems in irreversible thermodynamics.

The speed of electromagnetic waves

In our previous chapters, we assumed that light is an electromagnetic wave and that its speed does not depend on the frame of reference one employs. Let us now prove that this indeed is true. We begin with the free-space Maxwell equations:

$$\nabla \cdot \vec{E} = \frac{\rho}{\varepsilon_0}$$
$$\nabla \cdot \vec{B} = 0$$
$$\nabla \times \vec{E} = -\frac{\partial \vec{B}}{\partial t}$$
$$\nabla \times \vec{B} = \mu_0 \vec{J} + \varepsilon_0 \mu_0 \frac{\partial \vec{E}}{\partial t}.$$

Here, **E** and **B** are the electric and magnetic fields, ρ and **J** are the charge and the current densities, and ε_0 and μ_0 are the permittivity and permeability of the free space, respectively. The Maxwell equations give rise to solutions of wave-like behavior. To see the electromagnetic wave solution, we shall use a rather simple situation: the solution of Maxwell's equations in a region of space where there are no charges. Then, the Maxwell equations reduce to:

$$\nabla \cdot \vec{E} = 0$$

$$\nabla \times \vec{E} = -\frac{\partial \vec{B}}{\partial t}$$

$$\nabla \cdot \vec{B} = 0$$

$$\nabla \times \vec{B} = \varepsilon_0 \mu_0 \frac{\partial \vec{E}}{\partial t}.$$

Note the symmetry between **E** and **B** in this situation. We have

$$\nabla \times \left(\nabla \times \vec{E}\right) = -\frac{\partial\left(\nabla \times \vec{B}\right)}{\partial t} = -\varepsilon_0 \mu_0 \frac{\partial^2 \vec{E}}{\partial t^2}.$$

Using the vector identity $\nabla \times (\nabla \times \vec{E}) = \nabla(\nabla \cdot \vec{E}) - \nabla^2 \vec{E}$ and the idea that $\nabla \cdot \vec{E} = 0$, we obtain

$$\nabla^2 \vec{E} - \varepsilon_0 \mu_0 \frac{\partial^2 \vec{E}}{\partial t^2} = 0.$$

To make the physical concepts clear, we shall confine ourselves to a space of only one dimension, namely x. Then the above equation reduces to

$$\frac{\partial^2 \vec{E}}{\partial x^2} - \varepsilon_0 \mu_0 \frac{\partial^2 \vec{E}}{\partial t^2} = 0. \tag{8.22}$$

The general solution to equation (8.22) is of the form

$$\vec{E} = \vec{E}_0 e^{i(kx - wt)}.$$

The physical solution is the real part of the above equation: $\vec{E} = \vec{E}_0 \cos(kx - wt)$.

Here, w is the angular frequency and k is the wave number. The simplest form of the second-order homogenous wave equation is given by

$$\frac{\partial^2 \vec{E}}{\partial t^2} - c^2 \frac{\partial^2 \vec{E}}{\partial x^2} = 0,$$

where c is the speed of the wave. Looking at equation (8.22), we can immediately say that

$$c = \frac{1}{\sqrt{\varepsilon_0 \mu_0}}. \tag{8.23}$$

Substituting the values in equation (8.23), we obtain $c \approx 3 \times 10^8$ ms^{-2}. These results can be generalized to include the x, y and z axes and the general solution is of the form

$$\vec{E} = \vec{E}_0 e^{i\left(\vec{k}\cdot\vec{r} - wt\right)}$$
$$\vec{B} = \vec{B}_0 e^{i\left(\vec{k}\cdot\vec{r} - wt\right)},$$

where **r** represents the unit vector in the direction of the propagation of the wave. The speed of an electromagnetic wave as given by equation (8.23) is rather a remarkable result. If Maxwell's equations are valid in the same form in all inertial frames, light must travel with the same speed in all such frames, whose value is given by equation (8.23). This is, as we have seen before, one of the postulates of the special theory of relativity.

8.9 Some interesting remarks about 'displacement current'

In this section, we shall present some common misconceptions that students have about Maxwell's equations. The displacement current term was the final keystone added by Maxwell to complete the classical theory of electromagnetism. With the addition of the term, he was able to show that light is an electromagnetic wave and that these waves travel at the speed c. However, these claims made by Maxwell were based purely on theory. Hertz only demonstrated the experimental evidence for electromagnetic radiation 24 years later.

1. Maxwell saw the term 'displacement current' as a 'kind of elastic yielding to the electrical forces'. It was then believed that ether filled the entire empty space around us. Maxwell saw no difference in the electrical properties of ether and normal dielectrics. When an electric field is applied to dielectrics, polarization takes place and this results in the creation of an electric displacement field D. As we have seen, this can be written as $\vec{D} = \varepsilon_0\vec{E} + \vec{P}$. Maxwell made no attempt to separate the real displacement of charges in dielectrics represented by **P** and electric field **E** because it was thought that ether exhibited the same properties as a dielectric. Hence, he called the term $\frac{\partial \varepsilon_0 \vec{E}}{\partial t}$ 'displacement current', although we now know that it is not exactly a 'current' in the true sense, and that ether does not exist. Unfortunately, we are stuck with the name.
2. Let us go back to the problem with Ampère's law. We stated at the beginning of this chapter that Ampère's law is only valid when electric field is stationary with time and that it can be made complete to include all the time-dependent effects of the electric field by adding a term called displacement current. In particular, the Maxwell–Ampère law states that

$$\nabla \times \vec{B} = \mu_0\left(\vec{J} + \vec{J}_d\right).$$

We applied this equation to a finite current-carrying wire and showed that the divergence of the curl of the magnetic field is zero. A question that now arises is: do all currents form closed loops as a result of the displacement current term? The answer is 'yes'. This can also be seen in the capacitor example. There was no current in the space between the capacitor plates, yet the introduction of displacement current closed the circuit and allowed us to apply Ampère's law correctly.

3. In this subsection, we shall see the importance of the choice of the gauge in understanding the 'reality' of the displacement current. Maxwell published most of his work on electromagnetic theory in three papers: 'On Faraday's lines of force' (1855–6), 'On physical lines of force' (1861–2) and 'A dynamical theory of electromagnetic field' (1864). In his first paper, Maxwell provided a mathematical structure for Faraday's lines of forces. Faraday employed the 'field concept' to explain electric and magnetic phenomena. However, these Faraday fields are another mathematically equivalent way of demonstrating the action-at-a-distance formalism.

   The displacement current term occurs in the second of the above-mentioned papers, where Maxwell added the term 'to correct the equations of electric currents for the effect due to the elasticity of the medium'. The medium here includes both the ether and the dielectric medium. However, we see no arguments based on symmetry. He also employed mechanical models to explain the equations. It must be noted that 'two major innovations in Maxwell's electromagnetic theory—the displacement current and the electromagnetic theory of light—received their initial formulations in the context of the molecular-vortex model' [4]. In his paper, Maxwell also mentioned that 'the variation in displacement current is equivalent to a current, this current must be taken into account and added to the conduction current' for mathematical consistency. The important question that now arises is: how relevant is displacement current in contemporary physics? We have learnt that there is no such thing as ether filling empty space. So, what then is the physical meaning of displacement current? The key to understanding such questions lies in the use of a certain gauge.

   We have mentioned in this chapter that we are free to choose the divergence of the magnetic vector potential **A**. And, consequently, the two most widely used gauges are the Coulomb gauge and the Lorentz or retarded gauge. Maxwell insisted on the equivalence (and equivalent reality) of displacement current with conduction current, because of his consistent choice of the Coulomb gauge. Maxwell's equations in vacuum are

$$\nabla \cdot \vec{E} = \frac{\rho}{\varepsilon_0}$$

$$\nabla \cdot \vec{B} = 0$$

$$\nabla \times \vec{E} = -\frac{\partial \vec{B}}{\partial t}$$

$$\nabla \times \vec{B} = \mu_0 \vec{J} + \mu_0 \varepsilon_0 \frac{\partial \vec{E}}{\partial t}.$$

Using the scalar and the vector potential notation, we have that

$$\vec{E} = -\nabla\phi - \frac{\partial \vec{A}}{\partial t} \tag{8.24}$$

$$\vec{B} = \nabla \times \vec{A}. \tag{8.25}$$

Taking the divergence of equation (8.24) and the curl of equation (8.25), we have that

$$-\nabla^2\phi - \frac{\partial}{\partial t}\left(\nabla \cdot \vec{A}\right) = \frac{\rho}{\varepsilon_0} \tag{8.26}$$

$$-\nabla^2\vec{A} + \nabla\left(\nabla \cdot \vec{A}\right) = \mu_0\left[\vec{J} + \varepsilon_0\frac{\partial \vec{E}}{\partial t}\right]. \tag{8.27}$$

Now, we shall employ the two gauges and see what equations (8.26) and (8.27) tell us about the reality of displacement current.

Coulomb gauge

For the Coulomb gauge, we have that $\nabla \cdot \vec{A} = 0$. Then the equations for scalar and vector potentials are

$$\nabla^2\phi = -\frac{\rho}{\varepsilon_0} \tag{8.28}$$

$$\nabla^2\vec{A} = -\mu_0\left[\vec{J} + \varepsilon_0\frac{\partial \vec{E}}{\partial t}\right]. \tag{8.29}$$

Equation (8.28) is the Poisson equation and its solution is the so-called the instantaneous potential. Equation (8.29) has a similar form to the Poisson equation and its source term is the sum of the conduction current and the displacement current. It is in this sense that Maxwell described the displacement current as being 'electromagnetically equivalent' to the conduction current. Hence, in the Coulomb gauge, displacement current produces magnetic field, just like conduction current.

Lorentz gauge

For the Lorentz gauge, however, we have that $\nabla \cdot \vec{A} = -\varepsilon_0\mu_0\frac{\partial \phi}{\partial t}$. Now, equation (8.26) for the scalar potential becomes

$$-\nabla^2\phi + \varepsilon_0\mu_0\frac{\partial \phi}{\partial t} = \frac{\rho}{\varepsilon_0}. \tag{8.30}$$

This is a three-dimensional wave equation in which the information propagates at the speed of light c. Next, equation (8.27) becomes

$$-\nabla^2\vec{A} + \nabla\left(-\varepsilon_0\mu_0\frac{\partial\phi}{\partial t}\right) = \mu_0\left[\vec{J} + \varepsilon_0\mu_0\frac{\partial\vec{E}}{\partial t}\right]. \quad (8.31)$$

Using equation (8.30), equation (8.31) reduces to

$$-\nabla^2\vec{A} + \varepsilon_0\mu_0\frac{\partial^2\vec{A}}{\partial t^2} = \mu_0\vec{J}. \quad (8.32)$$

It is apparent from equation (8.32) that the only source term is the true conduction current **J**. The displacement current term does not appear here. Hence, if retardation effects are considered by means of the Lorentz gauge, the equations show more clearly that the true source of magnetic fields is true currents. However, if we use the Coulomb gauge, the displacement current term must be included in order for the theory to make mathematical sense. It should also be carefully noted that the choice of gauge is just a matter of convenience and that it is incorrect to state that the Coulomb gauge is wrong. The French physicist and mathematician, Henri Poincaré [5], stated in 1894 that

*In calculating **A** Maxwell takes into account the currents of conduction and those of displacement; and he supposes that attraction takes place according to Newton's laws, i.e. instantaneously. But, in calculating* [the retarded potential] *on the contrary we take account only of conduction currents and we suppose that the attraction is propagated with the velocity of light* [...] *It is a matter of indifference whether we make this hypothesis* [of a propagation in finite non-zero time] *and consider only induction due to conduction currents, or like Maxwell, we retain the old law of induction and consider both conduction and the displacement currents.*

4. Here we look at displacement current and magnetic field.

The displacement current is defined as

$$\vec{J}_d = \varepsilon_0\frac{\partial\vec{E}}{\partial t}.$$

Taking the curl of the above equation and using Faraday's law, we obtain that

$$\nabla\times\vec{J}_d = \varepsilon_0\frac{\partial\left(\nabla\times\vec{E}\right)}{\partial t} = -\varepsilon_0\frac{\partial^2\vec{B}}{\partial t^2}.$$

Hence, for relatively slowly changing fields ('quasi-static fields'), we have that the curl of the displacement current is negligible. Thus, in the quasi-static limit, we have that

$$\nabla\times\vec{J}_d \approx 0.$$

We can assume $\mathbf{J}_d$ to be made up of the superposition of radial currents, with either positive or negative divergence. The magnetic field produced by any such combination is always zero for reasons of symmetry. Hence, we have proved that in the quasi-static limit, the magnetic field produced by the displacement current is negligible. This is why experimental demonstrations of the effects of displacement current had to wait for Hertz, who demonstrated them 24 years after Maxwell formulated his theory.

8.10 Poynting's theorem

The Poynting theorem relates the energy stored in the electromagnetic field to the work done on an electrically charged object, through energy flux. This theorem is analogous to the work–energy theorem in classical mechanics.

In order to derive this theorem let us consider a particle of charge q moving with a velocity **v** in the presence of an electric field **E** and a magnetic field **B**. It then experiences a force called Lorentz force, given as shown in equation (8.33):

$$\vec{F} = q\left[\vec{E} + \left(\vec{v} \times \vec{B}\right)\right]. \tag{8.33}$$

Thus Lorentz force is the combination of electric and magnetic force on a point charge due to electromagnetic fields acting simultaneously. In order for work to be done, the force has to act over a displacement of the charge, so in time dt the charge moves $\upsilon \cdot \mathrm{d}t$. As the magnetic force always acts perpendicularly to the velocity, it does no work, so all the work dW comes from the electric field given by equation (8.34):

$$\mathrm{d}W = \vec{F} \cdot \vec{v}\ \mathrm{d}t = q\vec{E} \cdot \vec{v}\ \mathrm{d}t. \tag{8.34}$$

If we now consider a continuous distribution of charge with density ρ, then we can find the work done as given by equation (8.34) on a volume element d^3r by replacing q with $\rho\mathrm{d}^3r$, as shown in equation (8.35):

$$\mathrm{d}W = \rho\mathrm{d}^3r\vec{E} \cdot \vec{v}\ \mathrm{d}t = q\vec{E} \cdot \vec{v}\ \mathrm{d}t. \tag{8.35}$$

Since the current density $\vec{J}$ is the charge density times the velocity, $\vec{J} = \rho\vec{v}$, by replacing $\rho\,\vec{v}$ with $\vec{J}$ in equation (8.35), we obtain

$$W = JE\mathrm{d}^3r \cdot \mathrm{d}t. \tag{8.36}$$

The rate at which work is done over a volume V is therefore given by equation (8.37)

$$\frac{\mathrm{d}W}{\mathrm{d}t} = \int_{\upsilon} JE\ \mathrm{d}^3r. \tag{8.37}$$

We can write this entirely in terms of **E** and **B** by using Maxwell–Ampère's equation, as shown in equation (8.38)

$$\vec{E} \cdot \vec{J} = \vec{E}\frac{\left(\nabla \times \vec{B}\right)}{\mu_0} - \varepsilon_0\vec{E}\frac{\partial\vec{E}}{\partial t}. \tag{8.38}$$

An identity from vector calculus says $[\nabla \cdot (\vec{E} \times \vec{B}) = \vec{B} \cdot (\nabla \times \vec{E}) - \vec{E} \cdot (\nabla \times \vec{B})]$.
On rearranging this identity, equation (8.39) is obtained

$$\frac{1}{\mu_0}\vec{E} \cdot \left(\nabla \times \vec{B}\right) = \frac{1}{\mu_0}\left[\vec{B} \cdot \left(\nabla \times \vec{E}\right) - \nabla \cdot \left(\vec{E} \times \vec{B}\right)\right]. \tag{8.39}$$

Using Maxwell's third equation (Faraday's law), we obtain

$$\frac{1}{\mu_0}\vec{E} \cdot \left(\nabla \times \vec{B}\right) = -\frac{1}{\mu_0}\left[\vec{B} \cdot \left(\frac{\partial \vec{B}}{\partial t}\right) + \nabla \cdot \left(\vec{E} \times \vec{B}\right)\right]. \tag{8.40}$$

For any vector field $\vec{A}$, we have

$$\frac{\partial}{\partial t}A^2 = \frac{\partial}{\partial t}\vec{A} \cdot A = 2\vec{A} \cdot \frac{\partial}{\partial t}\vec{A}. \tag{8.41}$$

From equations (8.40) and (8.41), we have

$$\frac{1}{\mu_0}\vec{E} \cdot \left(\nabla \times \vec{B}\right) = -\frac{1}{\mu_0}\left[\frac{1}{2} \cdot \left(\frac{\partial \vec{B}^2}{\partial t}\right) + \nabla \cdot \left(\vec{E} \times \vec{B}\right)\right]. \tag{8.42}$$

From equations (8.38) and (8.42), we obtain (8.43)

$$\frac{1}{\mu_0}\nabla \cdot \left(\vec{E} \times \vec{B}\right) = -\frac{1}{2\mu_0}\frac{\partial \vec{B}}{\partial t} - \frac{\varepsilon_0}{2}\frac{\partial \vec{E}^2}{\partial t} - \vec{E} \cdot \vec{J}. \tag{8.43}$$

As we know, the magnetic flux density $\vec{B} = \mu_0\vec{H}$, where μ_0 is the permeability of the air and $\vec{H}$ is the magnetic field density, and by replacing $\vec{B}$ with $\mu_0\vec{H}$ in equation (8.43), we obtain equation (8.44)

$$\nabla \cdot \left(\vec{E} \times \vec{H}\right) = -\frac{1}{2}\frac{\partial \mu_0\vec{H}^2}{\partial t} - \frac{1}{2}\frac{\partial \varepsilon_0\vec{E}^2}{\partial t} - \vec{E} \cdot \sigma\vec{E}. \tag{8.44}$$

Equation (8.44) can be rewritten in integral form, as given by equation (8.45)

$$\int_V \nabla \cdot \left(\vec{E} \times \vec{H}\right)\mathrm{d}V = -\int_V \frac{1}{2}\frac{\partial \mu_0\vec{H}^2}{\partial t}\mathrm{d}V - \int_V \frac{1}{2}\frac{\partial \varepsilon_0\vec{E}^2}{\partial t}\mathrm{d}V - \int_V \sigma\vec{E}^2\,\mathrm{d}V, \tag{8.45}$$

where $\mathrm{d}V$ is the boundary of volume V.
The surface integral form of equation (8.45) is given by equation (8.46)

$$\oint_S \left(\vec{E} \times \vec{H}\right)n \cdot \mathrm{d}s = -\int_V \frac{1}{2}\frac{\partial \mu_0\vec{H}^2}{\partial t}\mathrm{d}V - \int_V \frac{1}{2}\frac{\partial \varepsilon_0\vec{E}^2}{\partial t}\mathrm{d}V - \int_V \sigma\vec{E}^2\,\mathrm{d}V, \tag{8.46}$$

where the integral is now over the surface S enclosing the volume V.

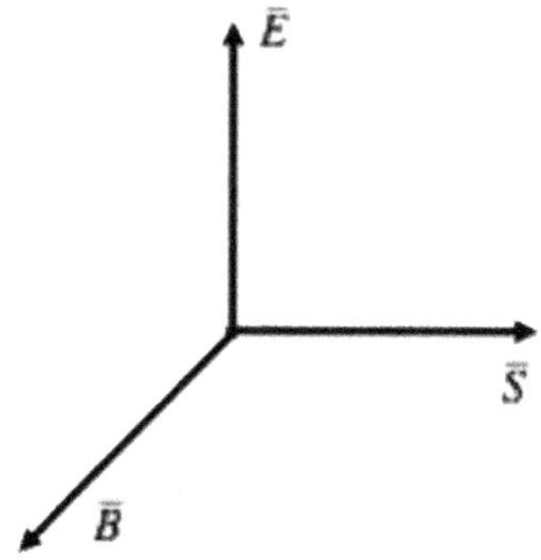

Figure 8.7. The direction of the electric and magnetic fields and the Poynting vector.

The RHS terms in equation (8.46):energy;

$\int_V \frac{1}{2}\frac{\partial \mu_0 \vec{H}^2}{\partial t} \mathrm{d}V$ represents the rate of change of stored magnetic energy;

$\int_V \frac{1}{2}\frac{\partial \varepsilon_0 \vec{E}^2}{\partial t} \mathrm{d}V$ represents the rate of change of stored electric energy;energy;

$\int_V \sigma \vec{E}^2 \, \mathrm{d}V$represents power dissipation as heat (Joule's law).

The LHS term of equation (8.46):energy;

$\oint_S (\vec{E} \times \vec{H})n \cdot \mathrm{d}s$ represents the power flowing out of the region.

The first two integrals of the RHS in (8.46) are the negative of the rate of change of the energy stored in the two fields. That is, as the fields do work on the charges they expend their energy (so it decreases, giving a negative derivative), but the work done on the charges is positive. The third term subtracts the work done by the field on the free electrical currents, which thereby exits from the electromagnetic energy as dissipation, heat, etc. The LHS integral is the rate at which energy flows across the surface S. By replacing $\vec{H}$ with $\frac{\vec{B}}{\mu_0}$ in the LHS term of equation (8.46), the vector field $\overline{S}$ that is known as the Poynting vector is given by

$$\vec{S} \equiv \frac{1}{\mu_0}\left(\vec{E} \times \vec{B}\right) \quad \text{(joules m s}^{-2}\text{)}. \tag{8.47}$$

The Poynting vector represents the rate per unit area at which energy crosses a surface. Thus $\overline{S}$ is the power per unit area and the direction is perpendicular to both the electric and magnetic fields, as shown in figure 8.7.

Conclusion

In this chapter, we have seen how the introduction of a term called 'displacement current' can solve the problems that occurred in applying Ampère's circuital law.

Maxwell first introduced this term while solving his magnetic vorticity equation. We also learnt the important Helmholtz theorem and understood the choice of gauge in representing scalar and vector potentials in electromagnetism. We then derived the Poynting vector, which was used in chapter 4 to find the direction of energy transfer in DC circuits.

Exercises

MCQs

1) In the Coulomb gauge, the magnetic field can be produced by:
   a) Moving charge
   b) A changing electric field
   c) Both
   d) None of the above
2) In the Lorentz gauge, the magnetic field is produced by:
   a) Moving charge
   b) A changing electric field
   c) Both
   d) None of the above
3) Displacement current goes through the plates of the capacitor and equals the current when the capacitor is being:
   a) Charged
   b) Discharged
   c) Both
   d) None of the above
4) If μ_0 represents the permeability and ε_0 represents the permittivity of free space, then the speed c of an electromagnetic wave in vacuum is given by:
   a) $c = \sqrt{\varepsilon_0 \mu_0}$
   b) $c = \dfrac{1}{\sqrt{\varepsilon_0 \mu_0}}$
   c) $c = \dfrac{1}{\varepsilon_0 \mu_0}$
   d) $c = \varepsilon_0 \mu_0$
5) Which of the following produces electromagnetic waves?
   a) A stationary charged particle
   b) A uniformly moving charged particle
   c) An accelerated charged particle
   d) None of the above

Problems

1) Calculate the energy $\mathrm{d}W$ passing during a time $\mathrm{d}t$ through a unit area perpendicular to the direction of the propagation of an electromagnetic wave.

2) The maximum electric field in a plane electromagnetic wave is 100 N/C. The wave is going in the x-direction and the electric field is in the y-direction. Find the maximum magnetic field in the wave and its direction.
3) Suppose that a parallel plate capacitor with a plate area A and a distance d between the plates is charged by passing a constant current I. Consider the surface S of the area A/k parallel to the plates and drawn symmetrically between them. Find the displacement current through the area S.
4) Protons with the same velocity v in the positive x-direction form a beam with current I. Find the magnitude and direction of the Poynting vector **S** outside the beat and at a distance d from the beam.
5) The energy crossing per unit area per unit time perpendicular to the direction of propagation is called the intensity of an electromagnetic wave. Derive an expression for the intensity of the electromagnetic wave if the electric and magnetic fields in the wave are described as

$$E = E_0 \sin \omega\left(t - \frac{x}{c}\right)$$

$$B = B_0 \sin \omega\left(t - \frac{x}{c}\right).$$

References

[1] Bork A M 1963 Maxwell, displacement current, and symmetry *Am. J. Phys.* **31** 854

[2] Charitat T and Graner F 2003 About the magnetic field of a finite wire *Eur. J. Phys.* **24** 267

[3] Reich G 2003 An alternative introduction to Maxwell's displacement current *Phys. Teach.* **51** 485

[4] Selvan K T 2009 A revisiting of scientific and philosophical perspectives on Maxwell's displacement current *IEEE Antennas Propag. Mag.* **51** 3

[5] O'Rahilly A 1938 *Electromagnetics* (London: Longman) p 176

IOP Publishing

The Foundations of Electric Circuit Theory

N R Sree Harsha, Anupama Prakash and D P Kothari

Chapter 9

Network theorems

9.1 Introduction

Now that we understand Maxwell's equations and Helmholtz's theorem, we are in a position to apply them to derive Kirchhoff's celebrated circuit theory laws. An electric circuit is defined as a path through which electrons can flow from voltage or current sources[1] due to electric field created by surface charges. Electric circuit theories enable us to determine the approximate relationships between electric currents and voltages at different points in the network. As a simplifying assumption, we shall treat the network components as lumped network parameters. Kirchhoff's laws are used to derive the current and voltage at different points in an electrical network. The German physicist G Kirchhoff first worked them out in 1847 [1]. They are divided into two laws, the Kirchhoff current law (KCL) and the Kirchhoff voltage law (KVL).

- The KCL states that the currents entering a node in a circuit are equal to the currents leaving the node.
- The KVL states that the directed sum of the potential differences encountered in a closed loop is equal to zero. In order to apply KVL, we first assume a direction of current. Next, we traverse the circuit along this direction of current, and if we encounter a negative terminal of a voltage source V, we write a $-V$ in the summation, and if we get a positive terminal, we write a $+V$. While traversing the loop, if a resistor R is encountered in which current i is flowing in the same direction as our own direction of traverse, we write $+iR$, otherwise we write $-iR$.

Let us see the application of these laws in the context of an example. Consider the circuit shown in figure 9.1. The voltage source E is ideal and the resistors R_1 and R_2 are ohmic and linear, meaning they obey Ohm's law $V = iR$.

[1] We shall use the terms 'circuit' and 'network' interchangeably.

doi:10.1088/978-0-7503-1266-0ch9

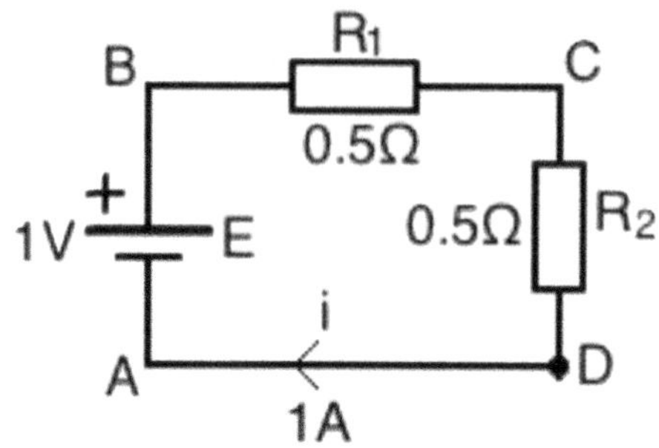

Figure 9.1. A simple circuit to demonstrate Kirchhoff's laws.

Kirchhoff's current law implies that the same current i flows through all the circuit elements. Let us assume that the direction of current is as shown in figure 9.1. Next, applying the KVL along the direction ABCD, the direction of the assumed current i, we obtain that

$$-E + iR_1 + iR_2 = 0$$

$$\Rightarrow i = \frac{E}{R_1 + R_2} = \frac{1}{0.5 + 0.5} = 1\ \mathrm{A}$$

We have seen in previous chapters that when there is a constant magnetic field or no magnetic field, we can write the electric field **E** present in space as

$$\nabla \times \vec{E} = 0. \tag{9.1}$$

Equation (9.1) entails the existence of a scalar potential, which we will call V. The existence of a scalar potential implies that the electric field **E** is a conserved quantity, like the gravitational field. Hence if we take a charged particle in a closed loop, the work done on it by the electric field is zero. This is essentially the proof of the KVL. Incidentally, we can apply the KVL along any closed path. There is no restriction that it has to be applied along the path that coincides with the circuit. However, the other paths that do not coincide with the circuit do not give any useful information about the lumped network elements.

Next, the KCL is just the applied version of the equation of the continuity of charge and current densities. In three-dimensional form, the KCL is represented by

$$\nabla \cdot \vec{J} = 0.$$

Although the above-mentioned proofs are correct, they are not very rigorous. In the next section we shall present a more rigorous proof of the Kirchhoff circuit laws (KCL and KVL).

9.2 Derivation of Kirchhoff's laws

We have seen how the steady state charge distribution creates an electric field inside and outside an electric circuit. It is immediately clear that such a charge distribution is unable to maintain an electric current in a DC circuit. The current in a DC circuit is the movement of charges in a closed conducting path. Since the electric field created by the stationary charge distribution on the conducting wires is a conservative field, it cannot do any work on the charges. And since current has to be driven against the resistance of the circuit, we conclude that there are non-conservative forces in a DC circuit that must drive the current against the resistance of the circuit. We saw in the previous chapter that Helmholtz's theorem states that any general vector field can be written as the sum of the conservative and non-conservative parts. Hence, we can write the electric field in the vicinity of a conducting DC circuit as the sum of the conservative field $\mathbf{E}_c$ and the non-conservative part $\mathbf{E}_{nc}$.

$$\vec{E} = \vec{E}_c + \vec{E}_{nc}.$$

While the conservative electric field is created by the stationary charge distribution on the circuit, the non-conservative part may be created by the chemical action inside a battery and only exists in the battery. Note that this non-conservative electric field is only used to model the chemical action of the battery on the charges inside the battery. Let us consider a simple DC circuit, as shown in figure 9.2.

A non-ideal battery is represented by an unusually large battery symbol and a resistor r between the positive and negative plates of the battery represents its internal resistance. The emf of the battery is taken to be E. A series combination of resistors R_1 and R_2 is connected to the battery, as shown in figure 9.2. The connecting wires that complete the DC circuit are considered to be ideal (zero resistivity wires). The switches S_1 and S_2 present on either side of the battery confine the charges present on the plates of the battery when they are open. The chemical action of the battery to maintain the charges on its plates is represented by a non-conservative field $\mathbf{E}_{nc}$, as shown in figure 9.2. The conservative electric field $\mathbf{E}_c$ created by the charges present on the plates of the battery is in the opposite direction to $\mathbf{E}_{nc}$, so the net electric field inside the battery is zero when the switches are open, and hence there is no current in the battery or in the DC circuit.

We now close switches S_1 and S_2, completing the circuit. The repelling charges on the plates of the battery are now distributed in the entire circuit, creating the electric

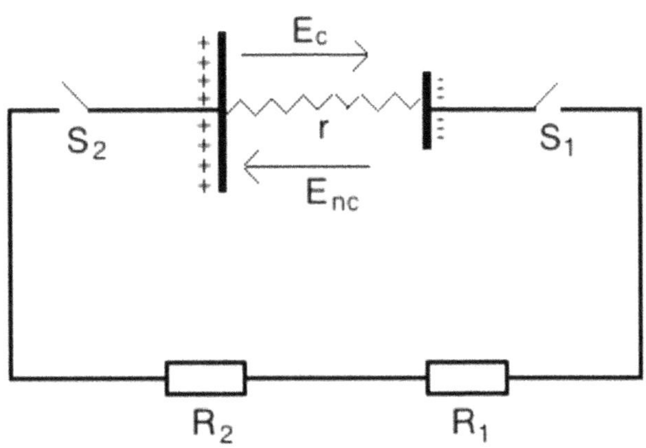

Figure 9.2. A simple DC circuit to prove Kirchhoff's laws.

fields in and around the circuit elements. Thus, the conservative field $\mathbf{E}_c$ decreases, while the non-conservative field (which is due to the chemical action of the battery) $\mathbf{E}_{nc}$ remains constant. Hence, there is now a net electric field $\mathbf{E}$ created inside the battery, as shown in figure 9.3.

We also have that the net electric field inside the battery is

$$\vec{E} = \vec{E}_c + \vec{E}_{nc}.$$

The electric fields $\mathbf{E}_1$ and $\mathbf{E}_2$ inside resistors R_1 and R_2 are shown in figure 9.3 and since they are created by the charge distribution present on the resistors, they are conservative fields. Next, suppose that an external agent moves a charge q quasi-statically (without acceleration) around the circuit starting at point e. The external agent does net zero work against the conservative forces. This implies that the applied force $\mathbf{F}$ from the external agent and the electric force due to the conservative charges must add to zero. Thus, we have that

$$\vec{F} + q\vec{E}_c = 0.$$

Thus the external agent must apply the force per unit charge given by

$$\frac{\vec{F}}{q} = -\vec{E}_c.$$

The total work w done by the external agent per unit charge in moving the charge in short displacements $\mathbf{dr}$ from e in the circuit in a counter-clockwise direction back to e must be zero, since it is only against conservative forces. Hence, we have that

$$w = \int_e^a -\vec{E}_c \cdot \mathrm{d}\vec{r} + \int_b^c -\vec{E}_c \cdot \mathrm{d}\vec{r} + \int_d^e -\vec{E}_c \cdot \mathrm{d}\vec{r} = 0. \tag{9.2}$$

Consider a general lumped circuit element, as shown in figure 9.4.

The current that flows through the element is I and the total electric field (conservative and non-conservative) inside the element is $\mathbf{E}$. The resistance of the

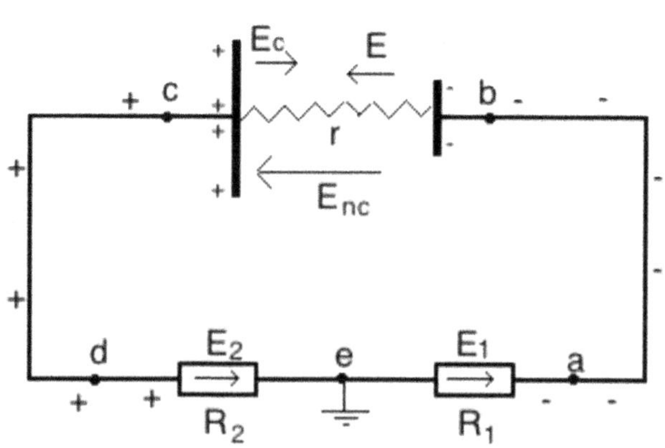

Figure 9.3. Switches S_1 and S_2 are closed, completing the path for current to flow.

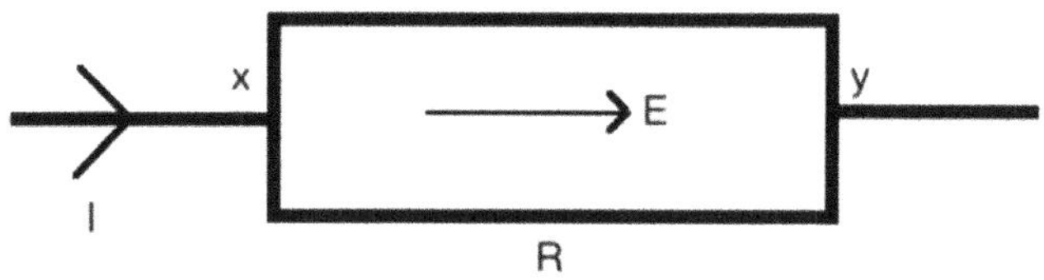

Figure 9.4. A general lumped circuit element.

element is R and the resistivity of the material of the element is ρ. Now, Ohm's law in its three-dimensional form relates these quantities as

$$\vec{J} = \frac{\vec{E}}{\rho}. \tag{9.3}$$

If the cross-sectional area of the element is A with length L, we have from equation (9.3) that

$$\int_x^y \vec{J} \cdot \mathrm{d}\vec{r} = \frac{1}{\rho}\int_x^y \vec{E} \cdot \mathrm{d}\vec{r}.$$

Now **J** is parallel to **dr**, hence the above integral reduces to

$$\vec{J}L = \frac{1}{\rho}\int_x^y \vec{E} \cdot \mathrm{d}\vec{r}.$$

Now multiplying on both sides by A/L, we obtain that

$$I = \frac{A}{\rho L}\int_x^y \vec{E} \cdot \mathrm{d}\vec{r}, \tag{9.4}$$

We have seen before that $\int_x^y \vec{E} \cdot \mathrm{d}\vec{r}$ represents the total work done w by the electric field **E** in moving the charges from x to y in the circuit element. Hence equation (9.4) reduces to

$$I = \frac{w}{R}, \tag{9.5}$$

where R is the resistance of the element. We have that $\mathbf{E} = \mathbf{E}_c = \mathbf{E}_1$ in resistor R_1, $\mathbf{E} = \mathbf{E}_c = \mathbf{E}_2$ in resistor R_2 and $\mathbf{E}_c = \mathbf{E} - \mathbf{E}_{nc}$ in the battery. We can now rewrite equation (9.2) as

$$-\int_e^a \vec{E}_1 \cdot \mathrm{d}\vec{r} + \int_b^c \vec{E}_{nc} \cdot \mathrm{d}\vec{r} - \int_b^c \vec{E} \cdot \mathrm{d}\vec{r} - \int_d^e \vec{E}_2 \cdot \mathrm{d}\vec{r} = 0. \tag{9.6}$$

Let us now assume that the current through resistors R_1 and R_2 is equal to I_1 and I_2, respectively. Let the current through the voltage source E be I_E.

As we have demonstrated, equation (9.6) can be written using equation (9.5) as

$$-I_1R_1 + E - I_Er - I_2R_2 = 0. \qquad (9.7)$$

We have taken $\int_b^c \vec{E}_{nc} \cdot d\vec{r}$ to be the emf E of the battery, consistent with our definition that emf is the work done by the *non-conservative electric field* in moving charge through the battery. We have yet to demonstrate that currents I_1, I_2 and I_E are equal. Consider a section S of the DC circuit shown in figure 9.5.

It is evident from figure 9.5 that the rate of change of the charge Q inside the section is given by

$$\frac{dQ}{dt} = I_E - I_2.$$

Suppose that we have $dQ/dt > 0$ at a particular instant. This implies that $I_E > I_2$. This means that the charge in section S increases, increasing the conservative electric field in both the battery and the resistor R_2. The increase of the conservative electric field decreases the net electric field inside the battery and hence decreases I_E. However, the increase in charge Q increases the conservative electric field inside the resistor and in turn increases I_2. Hence, we get the unstable situation of I_E decreasing and I_2 increasing until they become equal to each other.

Let us now suppose that $dQ/dt < 0$ at another instant. This implies that $I_2 > I_E$. Hence, the charge inside region S decreases and this in turn decreases the total conservative electric field inside the battery and the resistor. The decrease in the conservative electric field inside the battery increases the net electric field and current I_E, while the decrease in the conservative electric field in the resistor decreases current I_2. Hence, this condition forms an unstable situation in which I_2 decreases and I_E increases. This state of affairs continues until $I_2 = I_E$. Similar lines of thought can also be applied to resistor R_1 and hence we conclude that the only stable situation is one in which $I_2 = I_1 = I_E = I$. Thus equation (9.7) can be written as

$$-IR_1 + E - Ir - IR_2 = 0.$$

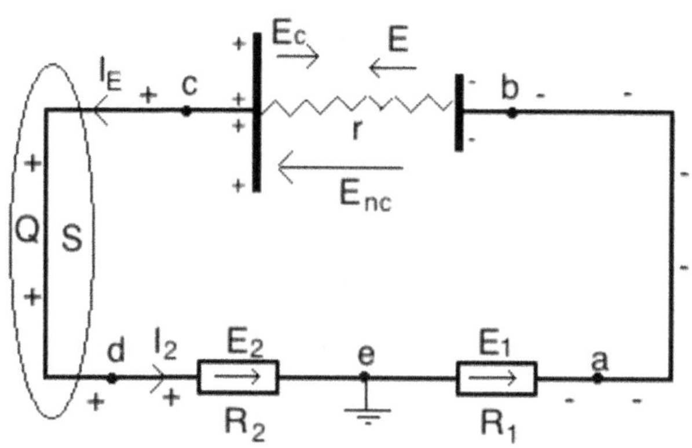

Figure 9.5. A section S is shown to demonstrate the proof of KCL.

This concludes the proof of Kirchhoff's laws. W R Moreau first proposed this proof in his brilliant paper [2] published in 1989. John R Carson also derived Kirchhoff's laws from retarded fields and potentials in 1927 [3]. Since he used the Lorentz gauge, this encompassed a wider range of circuit elements, such as capacitors and inductors.

9.3 The Newton of electricity

Before we discuss circuit analysis methods, we need to understand how voltage and the current source work. When Oersted first reported on his discovery of the action of electricity on a nearby magnet, André-Marie Ampère, a French physicist and mathematician, performed his experiments on current-carrying wires and tried to map the magnetic fields created around them. In the course of this mapping, he made a fundamental discovery, which later became the foundation of circuit analysis. At that time, it was generally believed that a voltaic pile acted like a battery of Leyden jars. But, contrary to this notion, Ampère discovered that the same current passed through the battery and the wire. Ampère thus found that the currents flowing through the voltaic battery and the wire are equal and formed the concept of the 'circuit', in which electric current is closed. In order to detect the electric currents in a wire, he also invented the so-called galvanometer. The contributions of Ampère were so important for the formulation of electromagnetism that Maxwell regarded him as 'the Newton of electricity' [4].

So, it became clear that a voltage source that is not closed cannot supply any current. In order to understand this more clearly, consider the circuit shown in figure 9.6.

An ideal voltage source E is connected to a series combination of ohmic resistors R_1 and R_2 and the zero potential is defined at point D in the circuit. With this point as a reference, we can define the potential at other points in the circuit. It is clear that the potential at point A, which we call V_A, is equal to zero, since it is connected to ground by a wire of zero resistance. Hence, we have

$$V_A = V_D = 0 \text{ V}$$

At the two terminals of the battery we must, by the very definition of a voltage source, have that

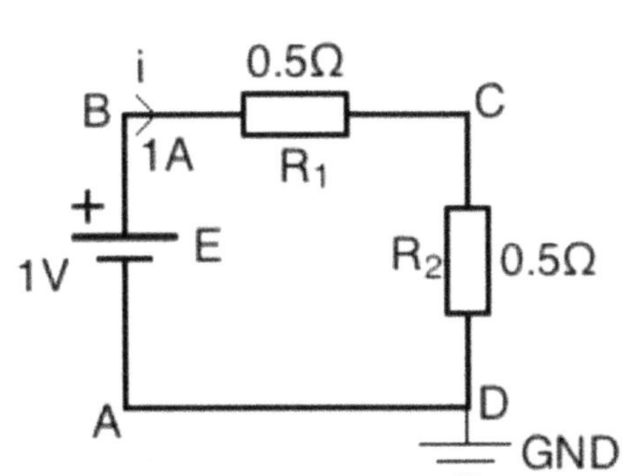

Figure. 9.6. A simple circuit to demonstrate the importance of the closed path.

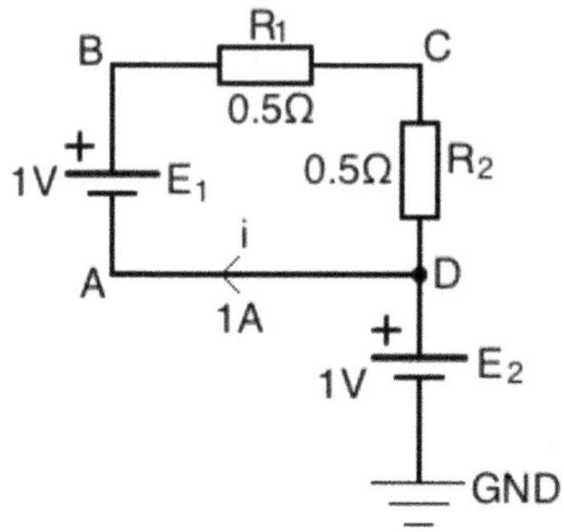

Figure 9.7. A simple circuit in which the ground is biased.

$$V_B - V_A = 1\text{ V}$$
$$\Rightarrow V_B = 1\text{ V}.$$

Applying Ohm's law to resistor R_1, we must have that

$$V_B - V_C = 1 \times 0.5\text{ V} = 0.5\text{ V}$$
$$\Rightarrow V_C = 0.5\text{ V}.$$

Next, applying Ohm's law to resistor R_2, we find that $V_D = 0$ V, as confirmed by our definition of the ground. Now, consider the circuit shown in figure 9.7.

We now have a voltage source E_2 connected to point D and the other terminal of the source is grounded. As discussed previously, the current through E_2 is zero because it does not have a return path for the current to flow. What, then, is the effect of source E_2? The voltage source E_2 elevates the potential of the entire circuit by 1 V. Hence, we now have that

$$V_A = V_D = 1\text{ V}$$
$$V_B = 2\text{ V} \quad \text{and}$$
$$V_C = 1.5\text{ V}$$

This can also be confirmed using the analysis presented for the previous case. Hence, the important point is that the voltage source does not provide any current when there is no return path for the currents to flow back into the negative terminal of the voltage source.

9.4 The concept of entropy in electrical circuits

Entropy is a crucial concept in thermodynamics and statistical physics. Thermodynamics originated in the mid-1800s as part of the effort to understand the gas laws. In 1865, Clausius coined the term *entropy* to define the quantity q/T.

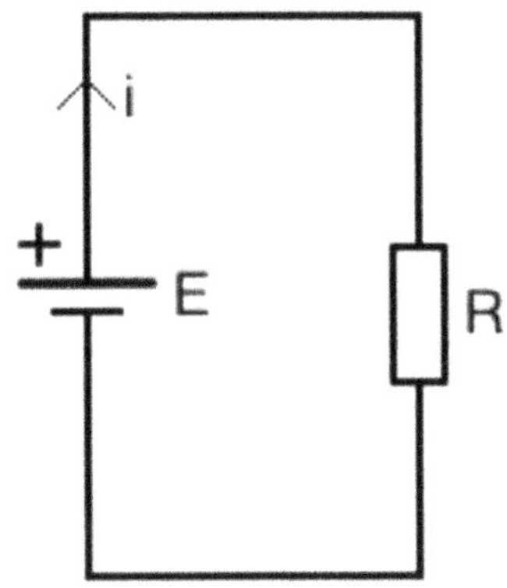

Figure 9.8. A simple circuit to determine the entropy produced by the resistor R.

Here, q is the heat content of the system and T is the temperature of the system. It was conceived to determine the equilibrium condition of the macroscopic states. Entropy, which is a function of the extensive parameters of the system, reaches its maximum value when the system attains an equilibrium state. In this section, we shall only be interested in defining the entropy in electric circuits. Consider the circuit shown in figure 9.8.

When a current i flows through the resistor R from the ideal battery E, entropy S is dispersed into the environment in the form of Joule heat. This conversion of the electrical energy into Joule heat in the ohmic resistor R is an irreversible process. Let us assume that the resistor operates at a constant temperature T. Then the rate of entropy production is given by

$$\frac{\mathrm{d}S}{\mathrm{d}t} = \frac{i^2 R}{T}.$$

Now, entropy is a first-order homogenous function of the extensive parameters of the system and hence it is an additive quantity for that system. If there are multiple resistors in the electric circuit, we should add the entropy produced in all resistors (labeled R_i with the currents through them labeled as i_i) for the total entropy S_T given to the environment by the electrical circuit. Hence, we have

$$\frac{\mathrm{d}S_T}{\mathrm{d}t} = \frac{\sum i_i^2 R_i}{T}.$$

Next, we shall see how the maximization of this rate of entropy production, with the energy constraint, gives us the solution for electric circuits.

9.5 Maximum entropy production principle

As we have seen, KVL represents the first law of thermodynamics in the electrical circuits. While there can be different distributions of currents in a network that satisfy the first law of thermodynamics, only certain unique values of the currents flow through the network components. This can be illustrated with the following example. Consider the circuit shown in figure 9.9.

KCL is valid at the node A. However, the current distribution is *not* correct, because the KVL is not valid in the circuit. However, the first law of thermodynamics is fully valid when applied to the entire circuit. The power drawn from the voltage source $V = 5$ V is given by

$$P_V = 5 \times 2 = 10 \text{ W}.$$

The power dissipated in the resistors is given by

$$P_R = 6 \times 1 + 4 \times 1 = 10 \text{ W}.$$

We can see that $P_V = P_R$. This implies that the currents distribute not only to make the first law of thermodynamics valid in the entire circuit, but also to make sure that KVL is valid in each loop. The reason why KVL is valid in each loop is not entirely clear from the thermodynamic point of view *per se*. In this section, we shall see that the currents distribute so as to maximize the rate of production of entropy. This is one of the many applications of the maximum entropy production principle (MEPP). Paško Županović *et al* first proposed this application of MEPP in electric circuits in 2004 [5], and argued that 'Kirchhoff's loop law follows from the principle of maximum overall entropy production in the network, assuming that the energy conservation is satisfied'. We shall illustrate this principle with the following simple example.

Consider the circuit shown in figure 9.10. Let the current from the ideal voltage source E be i and from the resistors R_2 and R_3 let it be i_1 and $i - i_1$, respectively. By making such an assumption, we are assuming that KCL is valid at node A. Now, MEPP states that currents i and i_1 are chosen by Nature so as to maximize the rate of production of entropy in the resistors, while satisfying the first law of thermodynamics in the entire circuit. Hence, we have

$$\begin{aligned} \text{maximize:} \quad & i^2R_1 + i_1^2R_2 + (i - i_1)^2R_3 \\ \text{subject to:} \quad & Ei = i^2R_1 + i_1^2R_2 + (i - i_1)^2R_3 \end{aligned}$$

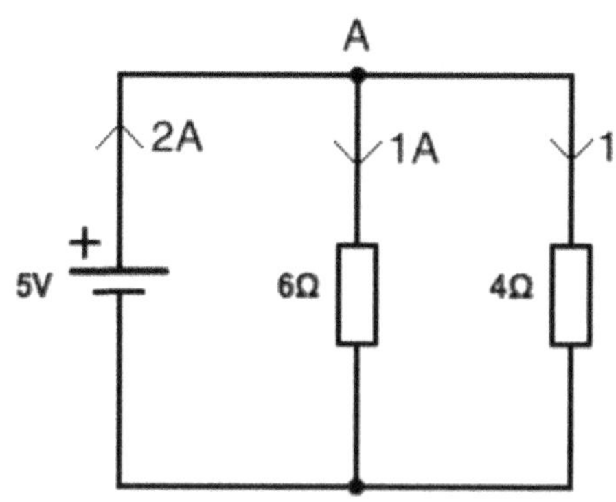

Figure 9.9. A circuit showing current distributions that obey the global conservation of energy. They do not, however, obey KVL along any loop.

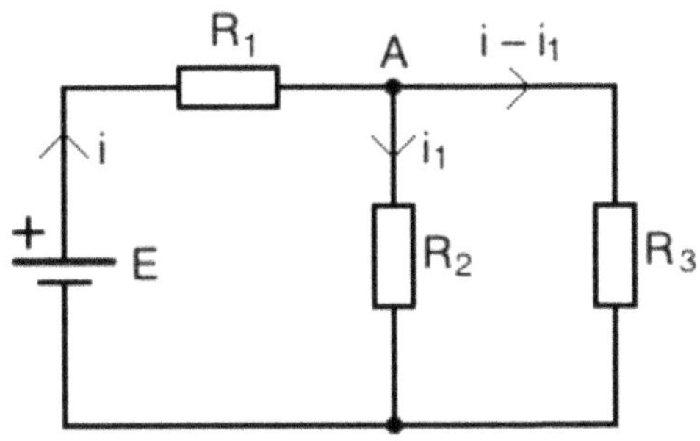

Figure 9.10. A simple circuit to demonstrate the application of the MEPP in electric circuits.

The technique for solving such problems is to introduce Lagrange multipliers, which were described in detail in chapter 1. We introduce the Lagrangian function Ψ as

$$\Psi = i^2R_1 + i_1^2R_2 + \left(i - i_1\right)^2 R_3 + \lambda\left(Ei - i^2R_1 - i_1^2R_2 - \left(i - i_1\right)^2 R_3\right).$$

In order to simplify the problem, we rewrite Ψ based on the global energy constraint as

$$\Psi = Ei + \lambda\left(Ei - i^2R_1 - i_1^2R_2 - \left(i - i_1\right)^2 R_3\right).$$

To define the extremum, we need that $\frac{\partial\Psi}{\partial i} = \frac{\partial\Psi}{\partial i_1} = 0.$

For $\frac{\partial\Psi}{\partial i} = 0$, we obtain that $\lambda \neq 0$. Next, defining $\frac{\partial\Psi}{\partial i_1} = 0$, we obtain that

$$\lambda\left(-2i_1R_2 + 2(i - i_1)R_3\right) = 0.$$

Hence, we obtain that

$$i_1 = \frac{iR_3}{R_2 + R_3}.$$

Substituting the above equation into the energy constraint, we obtain that

$$i = \frac{E\left(R_2 + R_3\right)}{R_1R_2 + R_2R_3 + R_3R_1}.$$

Hence, using the MEPP we have completely solved the currents in the different branches of the circuit. We have yet to prove that the rate of entropy production is a maximum. Let us now try to prove that for our example problem. As explained in the previous section, the rate of entropy production dS/dt is given by

$$\frac{\mathrm{d}S}{\mathrm{d}t} = \widehat{S} = \frac{P}{T}, \tag{9.8}$$

where P is the heat given to the environment by the electrical system at a temperature T. For our example, we have that

$$P = i^2R_1 + i_1^2R_2 + \left(i - i_1\right)^2 R_3 = Ei. \tag{9.9}$$

Next, from equation (9.8) we have that

$$\frac{\mathrm{d}P}{\mathrm{d}i_1} = E\frac{\mathrm{d}i}{\mathrm{d}i_1}, \tag{9.10}$$

which furnishes us with the relationship for the dependence of i on i_1. Differentiating P with respect to i_1 in equation (9.9) and using it in equation (9.10), we obtain that

$$\frac{\mathrm{d}i}{\mathrm{d}i_1} = \frac{i_1R_2 + i_1R_3 - iR_3}{E/2 - iR - iR_3 + i_1R_3}. \tag{9.11}$$

We also have from equation (9.8) that

$$\frac{\mathrm{d}\widehat{S}}{\mathrm{d}i_1} = \frac{1}{T}\frac{\mathrm{d}P}{\mathrm{d}i_1}. \tag{9.12}$$

Next, using equation (9.10) in equation (9.12), we obtain that

$$\frac{\mathrm{d}\widehat{S}}{\mathrm{d}i_1} = \frac{E}{T}\frac{\mathrm{d}i}{\mathrm{d}i_1}. \tag{9.13}$$

Substituting equation (9.11) into equation (9.13), we obtain that

$$\frac{\mathrm{d}\widehat{S}}{\mathrm{d}i_1} = \frac{E}{T}\left[\frac{i_1R_2 + i_1R_3 - iR_3}{E/2 - iR - iR_3 + i_1R_3}\right].$$

Now, for a steady state, we must have $\frac{\mathrm{d}\widehat{S}}{\mathrm{d}i_1} = 0.$

And $\frac{\mathrm{d}\widehat{S}}{\mathrm{d}i_1} = 0 \Rightarrow i_1R_2 + i_1R_3 - iR_3 = 0$. Hence, for a steady state we obtain

$$i_1 = \frac{iR_3}{R_2 + R_3}.$$

This is indeed the correct distribution of currents in the circuit. Next, at the steady state we have that

$$\left(\frac{\mathrm{d}^2\widehat{S}}{\mathrm{d}i_1^2}\right)\Bigg|_{\frac{\mathrm{d}\widehat{S}}{\mathrm{d}i_1}=0} = -\frac{2(R_2 + R_3)}{T} < 0.$$

Hence, we have proved for this particular example that the rate of production of entropy is a maximum. The demonstration of this example is based on T Christen's brilliant paper on the application of the MEPP in electrical systems [6]. For a general argument about the maximization of the rate of entropy production, the reader is advised to refer to the original papers by Paško Županović *et al* [5, 7].

9.6 Superposition theorem

The superposition principle states that the current through an element in a linear circuit is the algebraic sum of the currents through that element from each independent source acting alone. Now, linear circuits are circuits consisting of elements that obey the linearity property. If $g(x)$ is a general linear function of the variable x then, for linearity, we have that

$$g(x_1 + x_2) = g(x_1) + g(x_2).$$

Resistive circuits are linear. Consider the circuit shown in figure 9.11.

We can think of voltage sources E_1 and E_2 as the inputs and the current flowing through the load resistor R as the response or output. Analogous to the function $g(x)$ defined previously, we can write the current i as $i_{(E_1+E_2)}$ (i as a function of $(E_1 + E_2)$). Applying Kirchhoff's laws to the circuit, we obtain that

$$i_{(E_1+E_2)} = \frac{E_1 + E_2}{R} \tag{9.14a}$$

$$\Rightarrow i_{(E_1+E_2)} = \frac{E_1}{R} + \frac{E_2}{R} = i_{(E_1)} + i_{(E_2)}. \tag{9.14b}$$

With equations (9.14), we can write the response of the circuit shown in figure 9.11 as the sum of the responses of the two circuits shown in figure 9.12.

Hence, if there are multiple voltage sources in a linear circuit, the current through an element is the sum of the currents through the element from individual voltage sources acting alone. In the case of voltage sources, we should replace them by their internal resistance to determine the response due to other sources. This is the essence of the superposition theorem. Let us perform a similar demonstration with ideal current sources. Consider the circuit shown in figure 9.13.

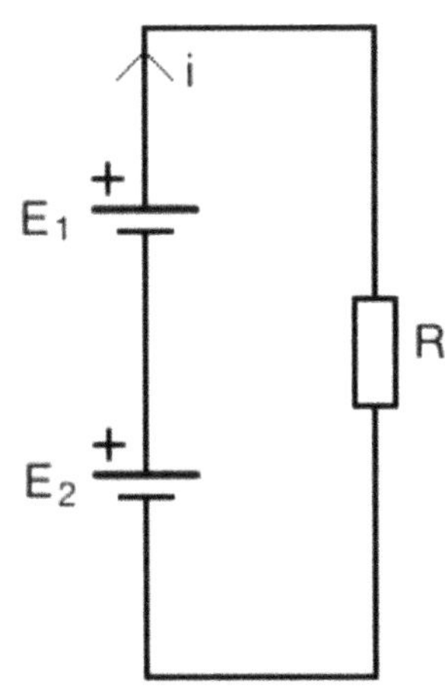

Figure 9.11. A circuit to demonstrate the superposition theorem in electric circuits.

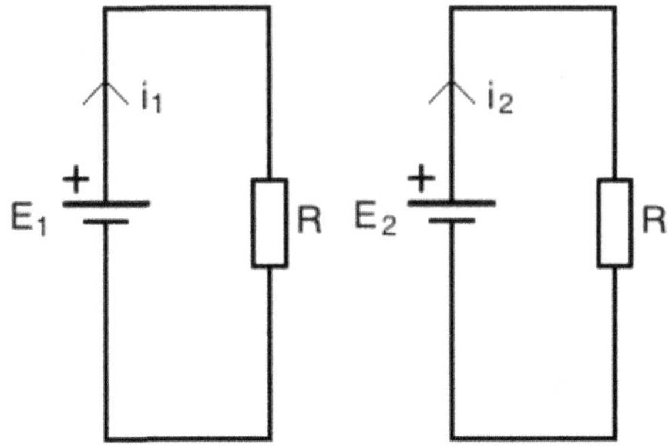

Figure 9.12. The sum of responses in these circuits is equal to the response in the circuit shown in figure 9.11.

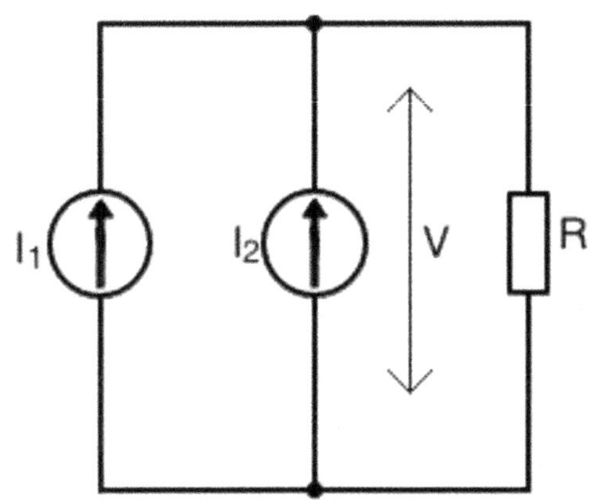

Figure 9.13. A circuit to demonstrate the application of superposition when current sources are present.

We now define the input as the current from the ideal current sources I_1 and I_2 and the voltage V across the load resistor as the response. Applying Kirchhoff's laws, we obtain that

$$V_{(I_1+I_2)} = (I_1 + I_2)R \tag{9.15a}$$

$$V_{(I_1+I_2)} = (I_1 + I_2)R = I_1R + I_2R = V_{(I_1)} + V_{(I_2)}. \tag{9.15b}$$

Hence, from equations (9.15) we can write the response of the circuit shown in figure 9.13 as the sum of the responses of the circuits shown in figure 9.14.

Hence, we arrive at another important result, i.e. that a current source, when the response of another current source is being calculated, should be replaced by an open circuit. This is fairly evident when we understand that the series combination of an ideal voltage source and an internal resistor of infinite values, as shown below, can be used to model an ideal current source. Consider the circuit shown in figure 9.15.

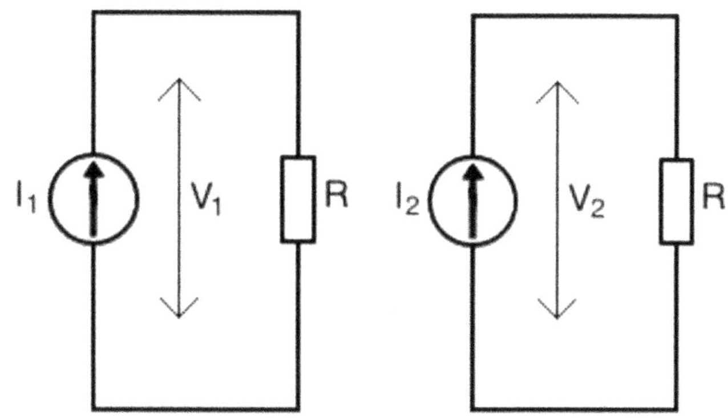

Figure 9.14. The sum of responses in these circuits is equal to the response in the circuit shown in figure 9.13.

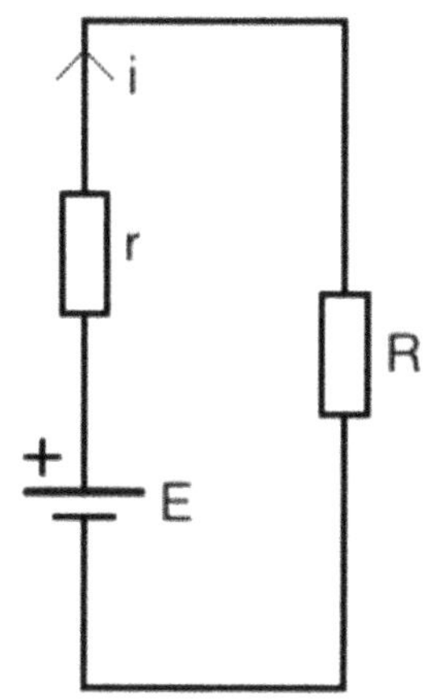

Figure 9.15. A simple circuit to demonstrate why a current source should be open-circuited when applying the superposition theorem.

Applying Kirchhoff's laws to the circuit shown in figure 9.15 gives us that

$$i = \frac{E}{r + R}$$

$$\Rightarrow i = \frac{E/r}{1 + R/r}.$$

If we now let the values of r and E tend to infinity, while keeping the ratio E/r constant, we can see that the current i does not depend on the load resistor R. This is basically the property of the ideal current source. Hence, in the superposition principle, we should replace the ideal current source with an open circuit, which is also essentially the series internal resistance of the ideal current source.

Example. Consider the circuit shown in figure 9.16. Using the superposition principle, find out which source provides the 1 A current to the load resistor.

Answer. We define the response in this particular problem as the current flowing through resistor R. Let us first consider the response from the ideal voltage source E.

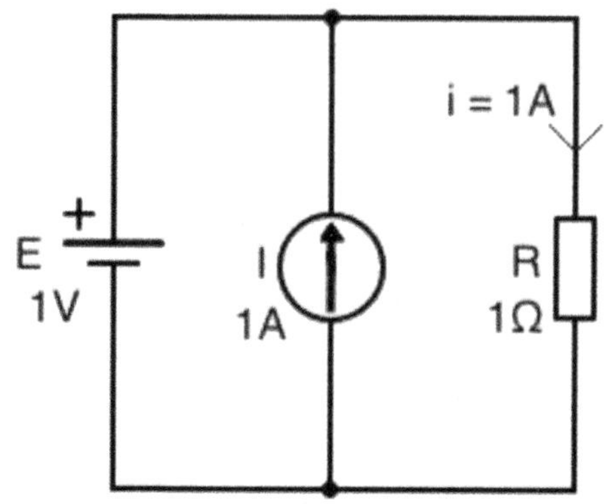

Figure 9.16. The figure used for the example problem.

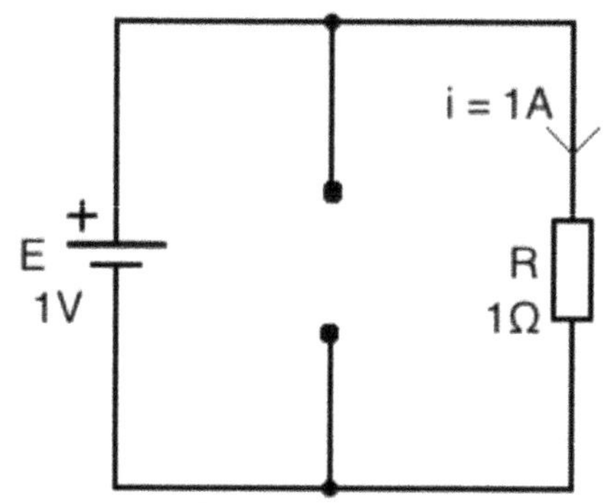

Figure 9.17. The current source open-circuited when considering the response from the voltage source *E*.

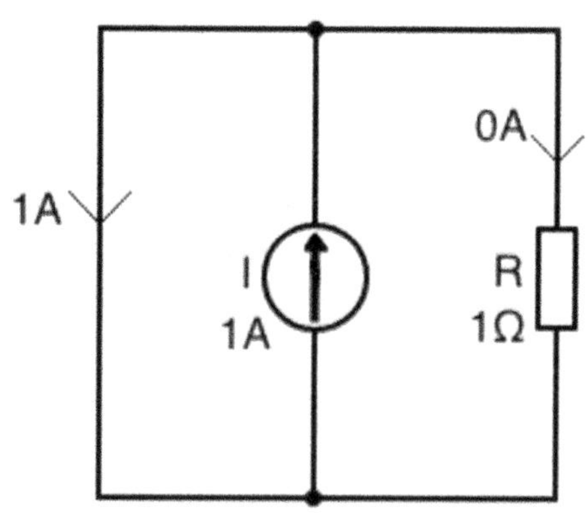

Figure 9.18. The voltage source short-circuited when considering the response from the current source.

We should open-circuit the current source *I* when considering the response from the voltage source. Hence, we have the circuit shown in figure 9.17.

We can see in figure 9.17 that the voltage source *E* provides the load current.

Next, we will consider the response from the current source *I*. The voltage source should be short-circuited and the entire current should flow through the zero resistance path, as shown in figure 9.18.

The final current through the voltage source E is the superposition of the current shown in the circuits in figures 9.17 and 9.18. Hence, we can see that the current through the voltage source cancels and no current is drawn from the voltage source. Thus, we conclude that the current source provides all the current to the resistor shown in figure 9.16.

9.7 Source transformation

Source transformation is the process of replacing the voltage source E in series with a resistor R by a current source I that is in parallel with the resistor R. We shall illustrate this method with a simple example. Consider the circuit shown in figure 9.19.

The current i through the load resistor is given by applying Kirchhoff's laws:

$$i = \frac{E}{r + R}.$$

Next, consider the circuit shown in figure 9.20.

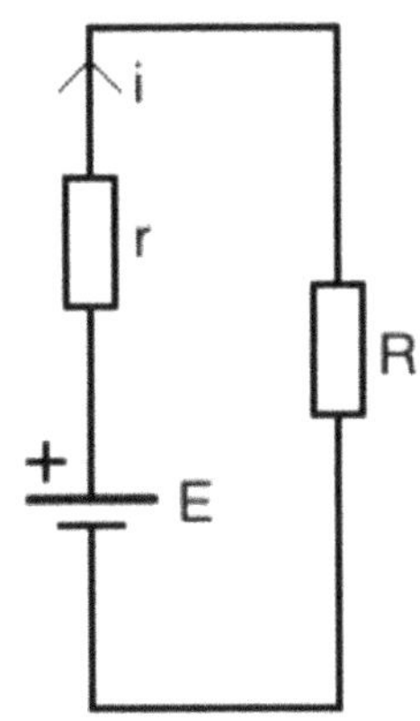

Figure 9.19. A circuit to demonstrate source transformation.

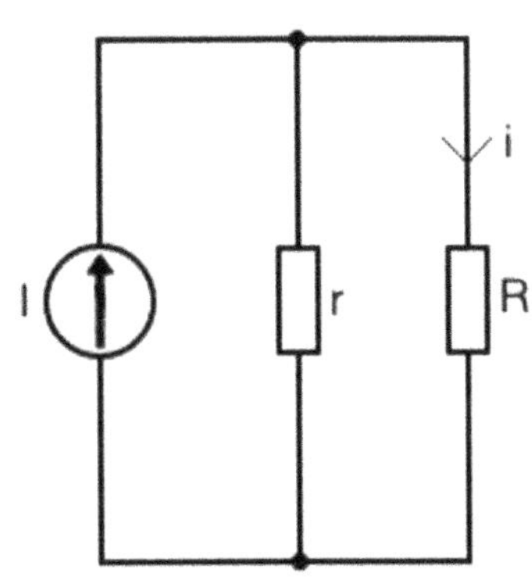

Figure 9.20. A circuit to demonstrate source transformation.

Applying Kirchhoff's laws gives us that

$$i = \frac{Ir}{r + R}.$$

Next, in order for the load currents through the load resistor R to be equal in figures 9.19 and 9.20, we have to set

$$E = Ir.$$

By doing this, we have transformed a voltage source into a current source without affecting the response at the load resistor. This is the essence of the source transformation principle.

9.8 Thevenin's theorem

In electrical engineering, we often encounter a situation in which the load resistance changes while the other parameters of the circuit remain constant. Then it is desirable to construct an equivalent model of the system in which only the load resistor is varied and calculate the new current. This is facilitated through Thevenin's theorem. The circuit shown in figure 9.21 illustrates this method.

Applying the superposition principle to the circuit gives us that

$$i = \frac{E + Ir}{r + R}.$$

Let us assume that the circuit components inside the 'black box' shown in figure 9.21 remain constant and that the load resistor R changes. It is inefficient to solve the entire circuit to calculate the current i through the resistor R. However, if we can write the black box as a series combination of a voltage source V_{Th} and a resistor R_{Th}, it is easier and more efficient to solve for the current i. We now give the rules for finding the values of V_{Th} and R_{Th}.

- The equivalent voltage V_{Th} is the voltage obtained at terminals A–B of the network in figure 9.21 with terminals A–B open-circuited.
- The equivalent resistance R_{Th} is the resistance that the circuit between terminals A and B would have if all the ideal voltage sources in the circuit were replaced by a short circuit and all the ideal current sources were replaced by an open circuit.

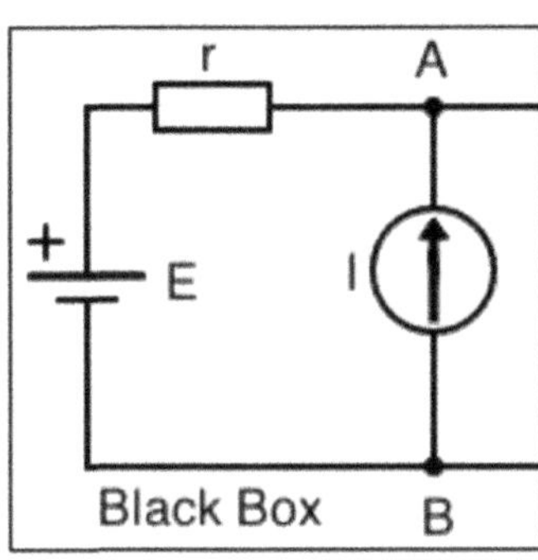

Figure 9.21. A circuit that produces the same load current i as the circuit shown in figure 9.25.

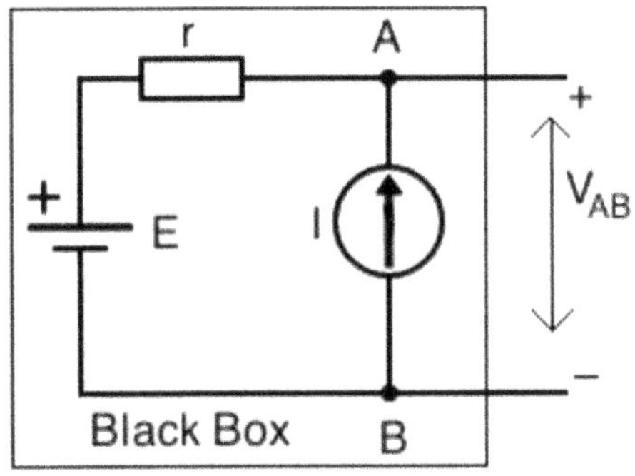

Figure 9.22. A circuit showing the black box in Thevenin's theorem.

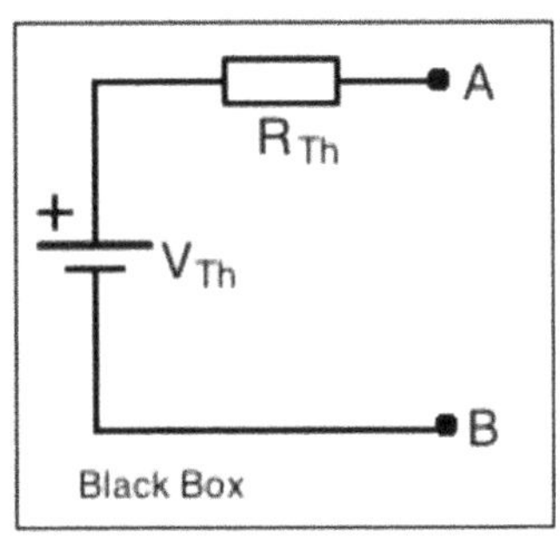

Figure 9.23. Another circuit that can produce the same output as the black box.

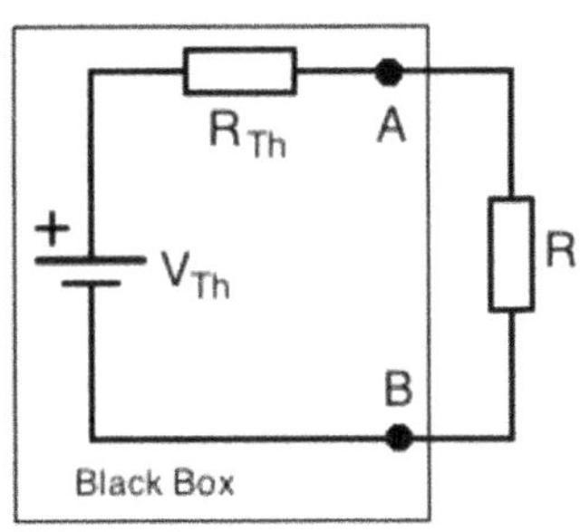

Figure 9.24. A circuit electrically equivalent to the original circuit shown in figure 9.21.

With these rules, let us now calculate the V_{Th} and R_{Th} for our example circuit in figure 9.21. For calculating the V_{Th}, consider the circuit shown in figure 9.22.

Applying the KVL along the outer loop, we obtain that

$$V_{\mathrm{Th}} = V_{AB} = E + Ir.$$

Next, R_{Th} is determined by replacing the voltage source E with a short circuit and the current source I with an open circuit and finding the resistance between the node points A and B in figure 9.21. We can easily see that $R_{\mathrm{Th}} = r$. Hence, the black box can be replaced by the circuit shown in figure 9.23.

Hence, we can now write the circuit shown in figure 9.21 as in figure 9.24.

With V_{Th} and R_{Th} derived as shown previously, we obtain the load current i that flows through the resistor R as

$$i = \frac{V_{Th}}{R_{Th} + R} = \frac{E + Ir}{r + R}.$$

Next, when the load resistor R changes value to a new value R_{new}, we can determine the current i_{new} that flows through it by using Thevenin's equivalent circuit and applying Kirchhoff's laws. We then obtain that

$$i_{new} = \frac{V_{Th}}{R_{Th} + R_{new}} = \frac{E + Ir}{r + R_{new}}.$$

Hence, we need not solve the entire circuit shown in figure 9.21 to determine the new value of the current. Thus, we have demonstrated Thevenin's theorem and its advantages.

9.9 Norton's theorem

Norton's theorem can be derived using Thevenin's theorem and the transformation of the sources. Once we have Thevenin's equivalent circuit, we can easily derive Norton's equivalent circuit by using the source transformation technique shown below. Consider the circuit shown in figure 9.25.

We can use source transformation to transform the voltage source V_{Th} in the circuit to a current source I_N, as shown in figure 9.26.

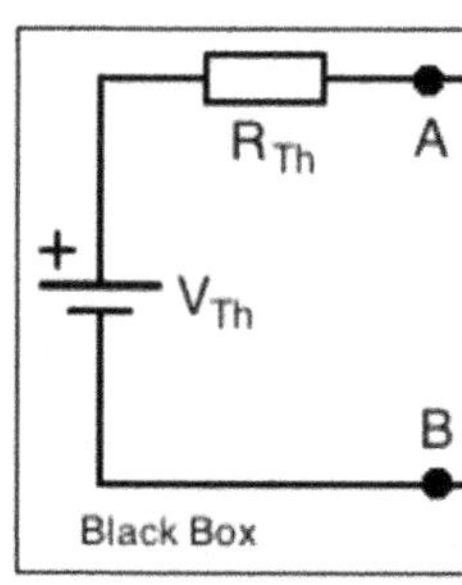

Figure 9.25. A circuit to demonstrate Norton's theorem.

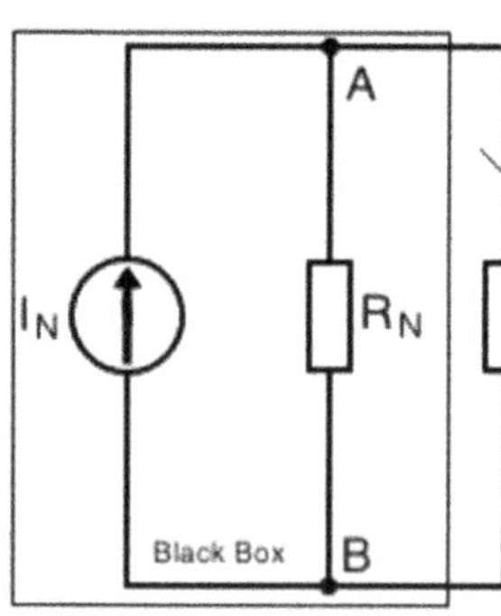

Figure 9.26. A circuit to demonstrate Norton's theorem.

As previously demonstrated, if we have $I_N = V_{Th}/R_{Th}$ with $R_N = R_{Th}$, we can see that the two circuits are completely equivalent as far as the load R is concerned. This is called Norton's theorem.

9.10 Tellegen's theorem in DC circuits

Tellegen's theorem is one of the most powerful theorems in circuit theory because it is applicable to all electrical circuits, whether they are linear or non-linear, time-invariant or time-dependent, hysteretic or non-hysteretic, or reciprocal or non-reciprocal [8]. A Dutch electrical engineer named D H Tellegen introduced this theorem in 1952.

Theorem

Suppose that there are n branches in an electrical circuit, which have $v_1, v_2, \ldots, v_n$ instantaneous voltages across them and $i_1, i_2, \ldots, i_n$ instantaneous currents through them. Then, the theorem [8] states that

$$\sum_{l=1}^{n} v_i i_i = 0,$$

where v_i and i_i are the voltage across and the current through the ith branch, respectively.

Let us illustrate this theorem with a simple example. Consider the circuit shown in figure 9.27.

Applying KVL along the outer loop, we obtain that

$$E = i_1 r + i_2 R. \qquad (9.16)$$

Next, applying KCL at node A, we obtain that

$$i_2 = I + i_1. \qquad (9.17)$$

Substituting this value of i_2 from equation (9.17) into equation (9.16), we obtain that

$$i_1 = \frac{E - IR}{r + R}. \qquad (9.18)$$

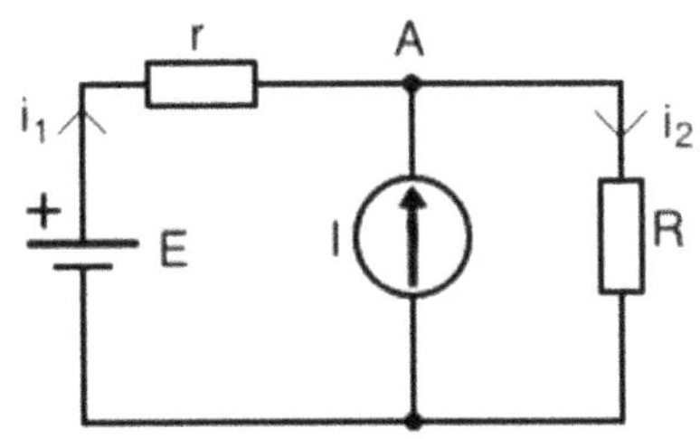

Figure 9.27. A circuit to demonstrate the application of Tellegen's theorem.

Substituting the value of i_1 from equation (9.17) into equation (9.18), we have that

$$i_2 = \frac{E + Ir}{r + R}.$$

Next, we have for Tellegen's theorem that

$$\sum_{i=1}^{3} v_i i_i = v_1 i_1 + v_2 i_2 + v_3 i_3$$

$$\Rightarrow v_1 i_1 + v_2 i_2 + v_3 i_3 = \left(-E + i_1 r\right) i_1 + \left(-i_2 R\right) I + \left(i_2 R\right) i_2$$

$$\Rightarrow v_1 i_1 + v_2 i_2 + v_3 i_3 = -i_1 i_2 R - I i_2 R + i_2^2 R$$

$$\Rightarrow v_1 i_1 + v_2 i_2 + v_3 i_3 = i_2[(i_2 - i_1)R - IR] = 0 \quad \text{(from equation (9.17))}.$$

Hence, we have demonstrated that for an electric circuit for which Kirchhoff's laws hold, we have $\sum_{i=1}^{n} v_i i_i = 0$. This is called Tellegen's theorem.

9.11 Some interesting remarks on Kirchhoff's laws

- Kirchhoff's voltage law cannot be applied when a changing magnetic flux passes through the circuit. In order to illustrate this, consider the situation shown in figure 9.28. A magnetic field **B** is present everywhere in space and increases with time in the direction shown in figure 9.28. Hence, we have a current in the circuit as shown in figure 9.28. For KCL to be valid, we need the same current i to flow through the two resistors R_1 and R_2.

  We use the integral form of Faraday's law along the loop ABCD, with S being the open surface attached to the loop, to obtain

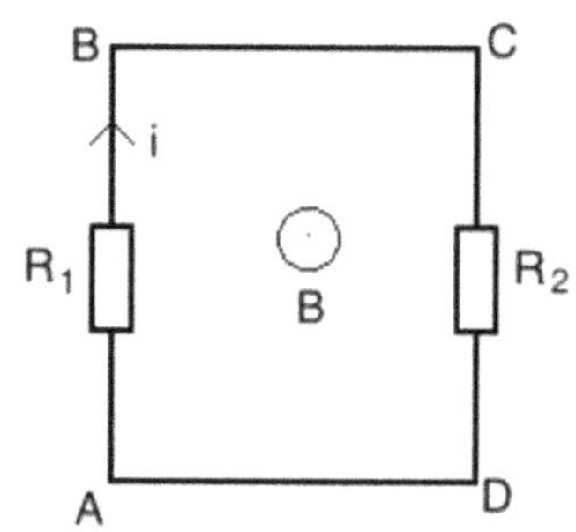

Figure 9.28. A circuit to show the non-conservative nature of the electric field produced by the changing magnetic field.

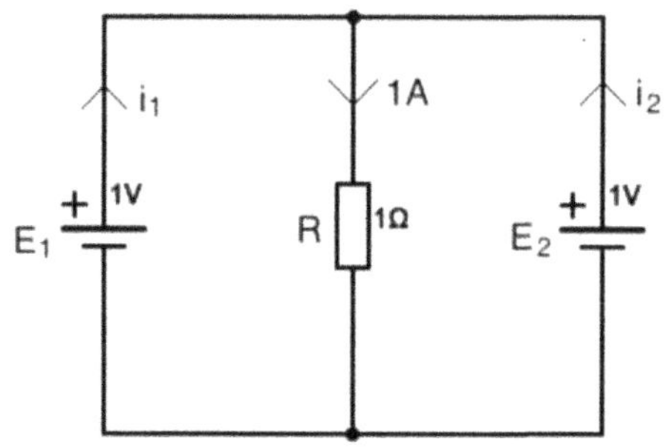

Figure 9.29. A circuit to show the failure of Kirchhoff's laws to determine i_1 and i_2.

$$\oint_{ABCD} \vec{E} \cdot d\vec{l} = -\frac{d}{dt}\left(\int_S \vec{B} \cdot d\vec{S}\right)$$

Hence, we see that the closed loop line integral of the electric field is no longer zero. This is because the electric field produced by changing the magnetic field is non-electrostatic and non-conservative in nature. Hence, KVL cannot be applied to the situation shown in figure 9.28. However, if we apply KVL naively, we obtain the absurd result that $(-iR_1) = iR_2$. Further details for this problem and the solution can be found in [9].

- Consider the circuit shown in figure 9.29. It demonstrates the complete breakdown of Kirchhoff's laws. In a recent paper [10], one of us proved that the currents i_1 and i_2 cannot be determined by any existing circuit analysis methods because the currents take the indeterminate 0/0 form. We shall not, however, go into this problem in detail here.

Exercises

MCQs

1) The standard derivation of Kirchhoff's voltage law assumes that the ___________ law of thermodynamics is satisfied in each loop.
   a. First
   b. Second
   c. Zeroth
   d. None of the above
2) The work done by an external agent in moving a positive test charge around a closed loop of a circuit against ___________ electric field/s is zero.
   a) Non-conservative
   b) Conservative
   c) The superposition of conservative and non-conservative
   d) None
3) Which of the following quantities do not change when a resistor connected to a battery is heated due to the current?
   a) Drift speed of the electrons

b) Resistivity
c) Resistance
d) None

4) The conversion of electrical energy into Joule heat in a resistor is:
   a. A reversible process
   b. An irreversible process
   c. Both
   d. None of the above
5) Assuming that only KCL is true, the stationary distribution of currents in a linear electric circuit is governed by the principle of ____________.
   a. Maximum entropy
   b. Maximum entropy production
   c. Minimum entropy
   d. The second law of thermodynamics

Problems

1) A wire of resistance 100 Ω is bent to make a complete circle. Find the resistance between two diametrically opposite points.
2) Find the current through the resistor R_2 in the circuit shown in figure M9.1 using the superposition theorem. Also find out which voltage source supplies current to resistor R_2.

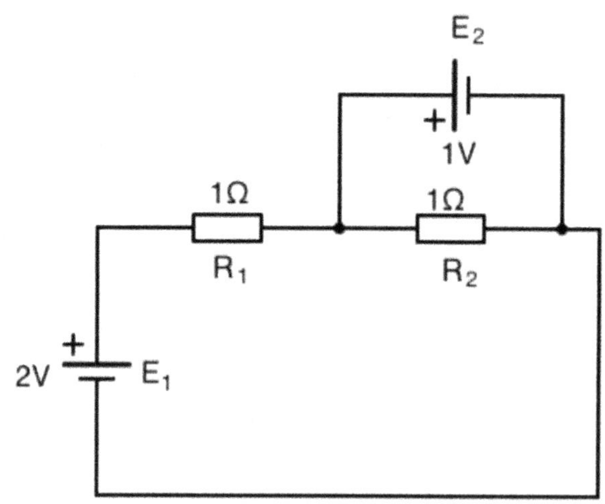

Figure M9.1.

3) Consider the circuit shown in figure M9.2. A current of 1 A flows through resistor *R*. Consider two points *a* and *b* on the ideal wire. The potential difference between *a* and *b* is zero as the wire is ideal and has no resistance. How can the current flow between the two points *a* and *b* if the potential difference is zero? Justify your answer.

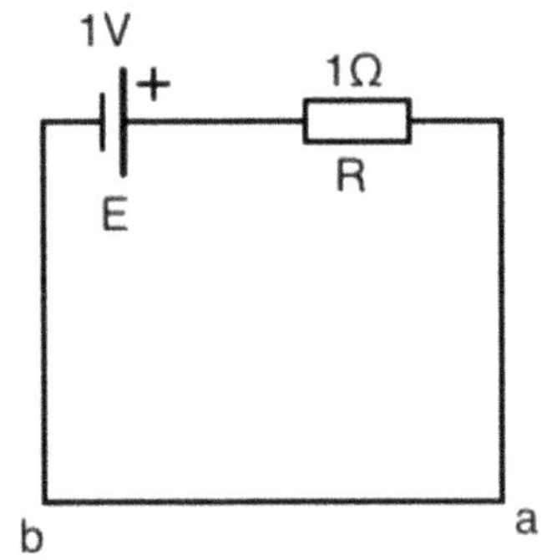

Figure M9.2.

4) Consider the circuit shown in figure M9.3. Find the product of the ambient temperature of the resistors and the rate of the entropy produced in the circuit.

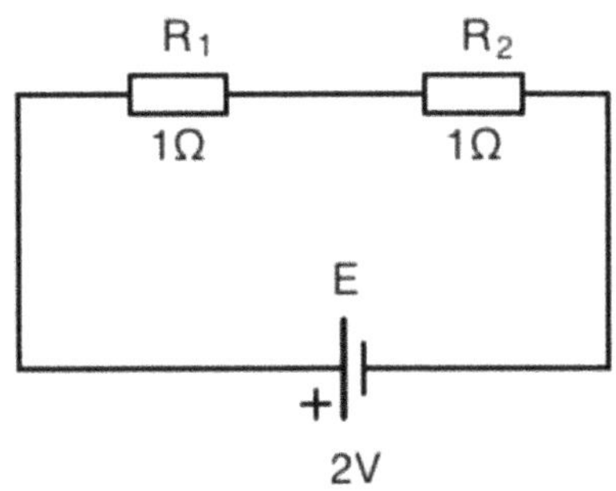

Figure M9.3.

5) For a circuit consisting of only a single loop, prove that the validity of Tellegen's theorem reduces to the validity of Ohm's law.

References

[1] Kirchhoff G 1847 *Ann. Phys. Chem.* **72** 497

[2] Moreau W R 1989 Charge distributions on DC circuits and Kirchhoff's laws *Eur. J. Phys.* **10** 286

[3] Carson J R 1927 Electromagnetic theory and the foundations of electric circuit theory *Bell Syst. Tech.* J **6** 1

[4] Darrigol O 2000 *Electrodynamics from Ampère to Einstein* (Oxford: Oxford University Press)

[5] Županović P *et al* 2004 Kirchhoff's loop law and the maximum entropy production principle *Phys. Rev.* E **70** 056108

[6] Christen T 2006 Application of the maximum entropy production principle to electrical systems *J. Phys. D: Appl. Phys.* **39** 4497

[7] Bortić S *et al* 2005 Is the stationary current distribution in a linear planar electric network determined by the principle of maximum entropy production? *Croatica Chemica Acta* **78** 181–4

[8] Penfield P *et al* 1970 A generalized form of Tellegen's theorem *IEEE Trans. Circuit Theor.* **CT-17** 302–5
[9] Romer R H 1982 What do 'voltmeters' measure? Faraday's law in a multiply connected region *Am. J. Phys.* **50** 1089
[10] Sree Harsha N R 2016 A singularity in Kirchhoff's circuit equations *Eur. J. Phys.* **37** 055801

IOP Publishing

The Foundations of Electric Circuit Theory

N R Sree Harsha, Anupama Prakash and D P Kothari

Chapter 10

Solutions-manual

Chapter 2

MCQs:

1) A fundamental particle or an elementary particle is a particle that is not made up of smaller particles. What are the fundamental particles of an atom?
   a) Quarks, gluons and electrons
   b) Protons, neutrons and electrons
   c) Protons, neutrons and electrons
   d) None of the above
   Answer: option (a).
2) The work done in moving a charge q along a circle in the presence of another charge Q is
   a) Zero
   b) Positive
   c) Negative
   d) Dependent on the radius of the circle
   Answer: option (a).
3) Two metallic spheres A and B of equal mass m are charged by the method of induction. In this process, suppose that A acquires positive charge and B acquires an equal amount of negative charge. Suppose that the masses of the spheres A and B after the process of induction are m_A and m_B, respectively. Which of the following is true?
   a) $m_A = m_B = m$
   b) $m_A < m_B = m$
   c) $m_A < m$ and $m_B > m$
   d) $m_A > m$ and $m_B < m$
   Answer: option (c).
4) Suppose that cotton is rubbed with amber to create frictional electrification. According to the triboelectric series, which of them will acquire a positive charge?
   Answer: cotton.

doi:10.1088/978-0-7503-1266-0ch10

5) We have seen that a fundamental particle or an elementary particle is a particle that is not made up of smaller particles. Which of the following is a fundamental particle?
   a) Atom
   b) Meson
   c) Proton
   d) Neutrino
   Answer: option (d).

Problems

(1) Suppose that the charge that flows through a circuit in time t and $t + \mathrm{d}t$ is given by

$$\mathrm{d}q = \mathrm{e}^{-\frac{t}{\tau}}\,\mathrm{d}t,$$

where τ is a constant. Find the total charge that flows through the circuit between $t = 0$ and $t = \tau$.

Solution:

The total charge Q is given by

$$Q = \int_o^\tau \mathrm{d}q = \int_0^\tau \mathrm{e}^{-\frac{t}{\tau}}\,\mathrm{d}t.$$

Upon evaluating the integral, we get that

$$Q = \left[\frac{\mathrm{e}^{-\frac{t}{\tau}}}{-\frac{1}{\tau}}\right]_0^\tau = \tau\left(1 - \frac{1}{e}\right).$$

(2) An electric field of magnitude E just prevents a water droplet of mass m from falling. Find the charge of the droplet.

Solution:

Let the charge of the droplet be q. We need to find out the forces acting on the water droplet. The two forces acting on the droplet are
   a) The electric force equal to qE
   b) The gravitational attraction equal to mg

For the electrical force to prevent the droplet from falling down, we must have that

$$qE = mg$$

$$\Rightarrow q = \frac{mg}{E}.$$

(3) When an object charge Q is divided into two objects, prove that the force of repulsion between the two will only be maximized if the total charge Q is divided equally between the two objects.

Solution:

Let us assume that one object receives a charge q. For the conservation of charge, we must require that the other object receive a charge $Q - q$. The magnitude of the force of repulsion F between them for a given distance r is given by

$$F = \frac{1}{4\pi\varepsilon_0}\frac{q(Q-q)}{r^2}.$$

For maximum, we need that $\frac{\mathrm{d}F}{\mathrm{d}q} = 0$. Hence, we have

$$\frac{\mathrm{d}F}{\mathrm{d}q} = \frac{1}{4\pi\varepsilon_0}\frac{Q-2q}{r^2} = 0$$

$$\Rightarrow q = \frac{Q}{2}.$$

Hence it is proved.

(4) Suppose we have two charged point particles with charge $-q$ and $2q$ and a distance r between them. Where can we place a third charge Q so that the total force experienced by the charge Q is zero?

Solution:

The two charged particles $-q$ and $2q$ are in a straight line, as shown in figure M2.1.

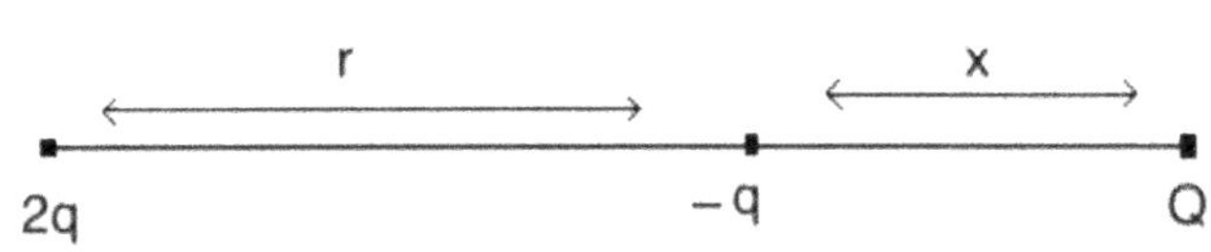

Figure M2.1.

Let the force due to charge $-q$ be F_1 and let the force due to charge $2q$ be F_2. We have that

$$|F_1| = \frac{Qq}{4\pi\varepsilon_0 x^2}.$$

We also have that

$$|F_2| = \frac{2Qq}{4\pi\varepsilon_0 (r+x)^2}.$$

For Q to experience zero net force, we have that

$$|F_2| = |F_1|.$$

This gives us that

$$(r + x)^2 = 2x^2.$$

Solving the equation, we have that either $x = -1.8r$ or $x = 3.8r$. Neglecting the negative value for distance, we have that $x = 3.8r$.

(5) The protons in a nucleus are separated by a distance of 1 fermi (1 fermi = 10^{-15} m). Find the electric force between two protons.

Solution:

$$F = 9 \times 10^9 \frac{1.6 \times 1.6 \times 10^{-38}}{10^{-30}},$$

$$F = 230.4 \text{ N}.$$

Chapter 3

MCQs:

1) Figure M3.1 shows some of the electric field lines corresponding to an electric field with magnitude E. What does the figure suggest?

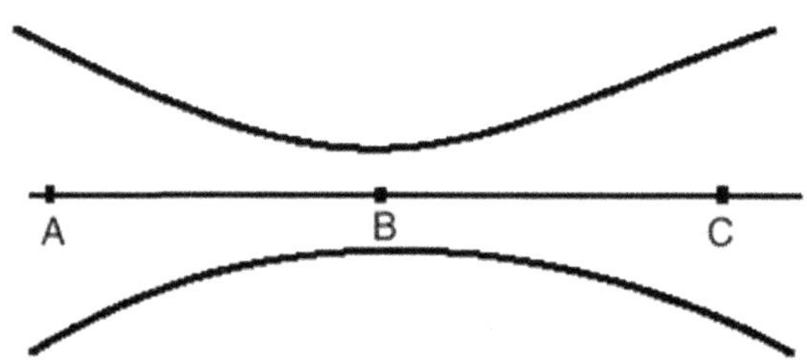

Figure M3.1.

a) $E_1 > E_2 > E_3$
b) $E_1 = E_3 < E_2$
c) $E_1 = E_2 = E_3$
d) $E_1 = E_3 > E_3$
Answer: option (b).

2) Suppose that a proton and an electron are placed in a uniform electric field. Which of the following is true?
a) The electric forces acting on them are equal
b) Their acceleration will be equal
c) The magnitude of the electric forces on them is equal
d) The magnitude of the acceleration will be equal
Answer: option (c).

3) The total force **F** experienced by a charged particle of charge q moving with a velocity **v** in an electric field **E** and a magnetic field **B** is given by:
   a) $\vec{F} = q(\vec{E} + \vec{B})$
   b) $\vec{F} = q(\vec{B} + \vec{v} \times \vec{E})$
   c) $\vec{F} = q(\vec{v} \times \vec{B} + \vec{E})$
   d) None of the above
   Answer: option (c).
4) A metallic particle with no net charge is placed near a finite metal plate carrying a negative charge. The electric force experienced by the particle is:
   a) Towards the plate
   b) Away from the plate
   c) Zero
   d) None of the above
   Answer: option (a).
5) A positive point charge q is brought near an isolated metal cube. Which of the following is true?
   a) The cube becomes positively charged
   b) The cube becomes negatively charged
   c) The interior remains charge-free but the surface obtains a non-uniform charge distribution
   d) None of the above
   Answer: option (c).

Problems

1) A uniform electric field of magnitude **E** exists in space in the x-direction. Calculate the flux of this field through a plane square of edge a placed in the yz plane.

Solution:

The flux is given by

$$\Phi = \int \vec{E} \cdot \mathrm{d}\vec{S}.$$

As the normal to the area points along the electric field, we have that the angle between the two vectors **E** and **dS** is zero. Hence, we have that

$$\Phi = Ea^2.$$

2) The intensity of an electric field depends on the x and y coordinates as shown:

$$\vec{E} = \frac{k\left(x\hat{i} + y\hat{j}\right)}{\left(x^2 + y^2\right)}$$

Here, k is a constant. Find the charge within a sphere of radius r with its center at its origin.

Solution:

From Gauss's theorem, we have that the required charge is equal to the flux of the electric field through this sphere multiplied by ε_0. We know that the electric field is axisymmetric. Hence, we can determine the flux through a cylinder of height $2r$ and it should be equal to the flux through the sphere. Hence, we have that

$$q = \varepsilon_0 \oint_S \vec{E} \cdot \mathrm{d}\vec{S} = \varepsilon_0 E_r S,$$

where $E_r = k/r$ and $S = 4\pi r^2$. Hence, $q = 4\pi\varepsilon_0 kr$.

3) A uniform electric field exists in space in the x-direction. Find the flux of this field through a cylindrical surface with the axis parallel to x-axis.

Solution:

The cylindrical surface has a normal that is perpendicular to the field. If we consider any small element on the surface, we know that the angle its area vector makes with respect to the electric field is 90°. Hence, the total flux of the electric field through the surface is zero.

4) A charge Q is uniformly distributed on a ring of radius R. A Gaussian sphere of radius R is formed with its center at the periphery of the ring, as shown in figure M3.2. Find out the flux of the electric field through the surface of the sphere.

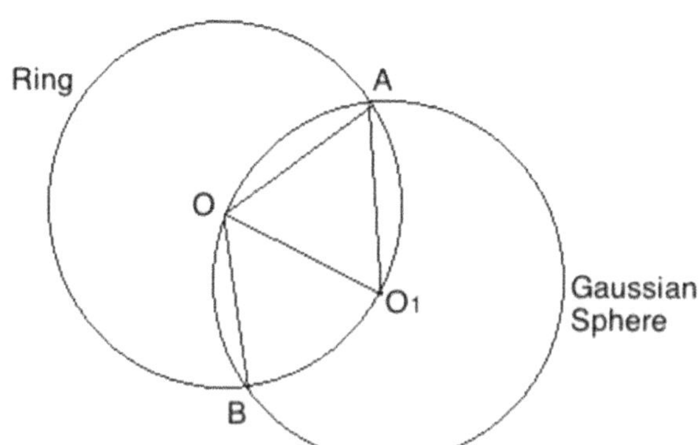

Figure M3.2.

Solution:

It can easily be seen from the figure that the triangle AOO_1 forms an equilateral triangle. Hence, the angle AOB is 120°. Since the ring subtends a total of 360°, we conclude that one third of the ring is in the Gaussian sphere. The flux through the closed surface is equal to the charge enclosed in the surface divided by ε_0 by Gauss's theorem. Hence, the flux of the electric field through the Gaussian surface is $Q/3\varepsilon_0$.

Chapter 4

MCQs:

1) When the separation between the charges is decreased, the electric potential energy of the charges:
   a) Increases
   b) Decreases
   c) Remains the same
   d) May increase or decrease
   Answer: option (d).
2) If a negative charge is shifted from a low potential region to a high potential region, the electric potential energy:
   a) Increases
   b) Decreases
   c) Remains the same
   d) May increase or decrease
   Answer: option (b).
3) The magnitude of the electric field and the electric potential at a point in space are E and V, respectively. Which of the following is true?
   a) If $E = 0$, V must be zero
   b) If $V = 0$, E must be zero
   c) If $E \neq 0$, V cannot be zero
   d) None of the above
   Answer: option (d).
4) Which of the following quantities depend on the choice of zero potential?
   a) Potential at a point
   b) Potential difference between two points
   c) Electric field at a point
   d) None
   Answer: option (a).
5) A capacitor of capacitance C is charged to a potential V. Suppose we have a closed surface S that encloses the capacitor. The flux of electric field through the surface S multiplied by ε_0 is equal to:
   a) CV
   b) $2CV$
   c) $CV/2$
   d) Zero
   Answer: (d).

Problems

1) Two charges 10 and 20 nC are placed at a distance of 2 m. Find the electric potential due to the pair at the middle of the line joining the two charges.

Solution:

We know that the potential due to point charge is $V = \dfrac{Q}{4\pi\varepsilon_0 r}$. The potential V_1 due to 10 nC charge is

$$V_1 = \frac{10 \times 10^{-9} \times 9 \times 10^{9}}{1} = 90 \text{ V}$$

Similarly, the potential V_2 due to 20 nC charge is 180 V. The total potential at the given point is 90 + 180 = 270 V.

2) Three charged particles that each have a charge of 10 nC are placed at the vertices of an equilateral triangle with a side of 10 cm. Find the work done by an external agent in pulling them apart to infinite separation.

Solution:

The potential energy of the system in the initial condition is

$$U = 3\left[\frac{qq}{4\pi\varepsilon_0 r}\right].$$

Upon substituting the values, we obtain that $U = 27\ \mu$J. When the charges are infinitely separated, the potential energy reduces to zero. Hence, the external agent must do total work of $-U$ to separate them.

3) Find the equivalent capacitance between the points A and B in the diagram shown in figure M4.1.

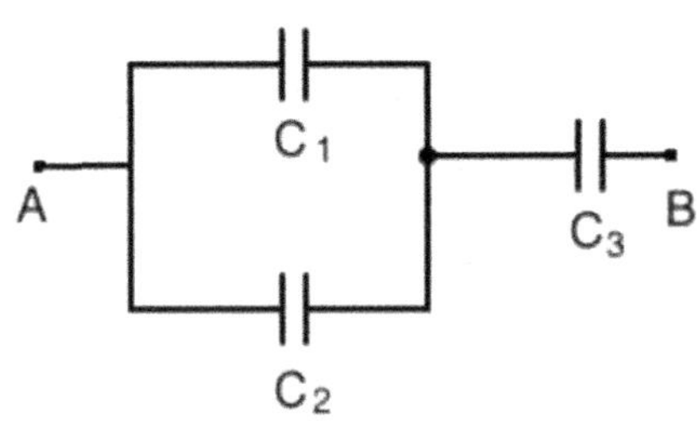

Figure M4.1.

Solution:

The capacitors C_1 and C_2 are connected in parallel. Hence, the equivalent capacitance is $C' = C_1 + C_2$. This equivalent capacitor and C_3 are connected in series. Hence the equivalent capacitance C_E between A and B is given by

$$\frac{1}{C_E} = \frac{1}{C'} + \frac{1}{C_3} = \frac{1}{C_1 + C_2} + \frac{1}{C_3}$$

$$\Rightarrow C_E = \frac{C_1C_3 + C_2C_3}{C_1 + C_2 + C_3}.$$

4) Calculate the charge on each capacitor shown in figure M4.2.

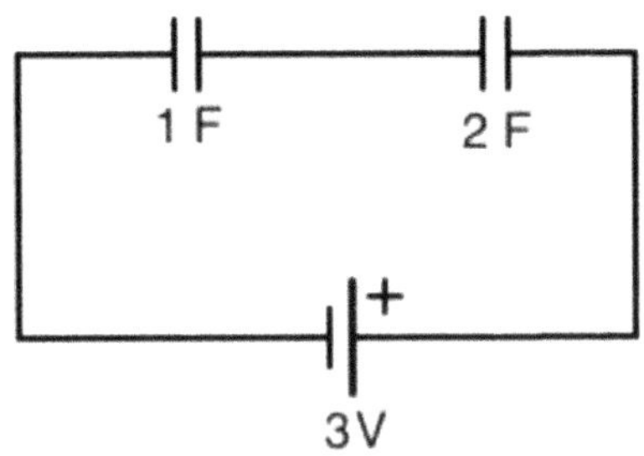

Figure M4.2.

Solution:

Since the two capacitors are connected in series, their equivalent capacitance is given by

$$C_E = \frac{C_1C_2}{C_1 + C_2} = \frac{2}{3}F.$$

The charge supplied by the battery is then given by $Q = CV$.

$$Q = \frac{2}{3}F \times 3V = 2C.$$

This charge supplied by the battery is the charge in each capacitor since they are connected in series.

5) Calculate the capacitance of a parallel plate capacitor with 2 × 2 square plates separated by a distance of 1 mm.

Solution:

The capacitance is given by $C = \frac{\varepsilon_0 A}{d} = \frac{8.85 \times 10^{-12}\ \mathrm{F\,m^{-1}} \times 400\ \mathrm{m^2}}{1 \times 10^{-3}\ \mathrm{m}} \approx 3.5 \times 10^{-6}$ F.

Chapter 5

MCQs:

1) If a constant force F acts on a particle, its acceleration will:
   a) Remain constant
   b) Gradually decrease
   c) Gradually increase
   d) None of the above
   Answer: option (b).

2) Which of the following quantities related to a particle has a finite upper limit?
   a) Kinetic energy
   b) Momentum
   c) Speed
   d) None of the above
   Answer: option (c).
3) In an electric circuit containing a battery, the negative charge inside the battery:
   a) Always goes from the positive terminal to the negative terminal
   b) May go from the negative terminal to the positive terminal
   c) Always goes from the negative terminal to the positive terminal
   d) None of the above
   Answer: option (b).
4) A resistor R is connected to an ideal battery. As the resistance of the resistor R increases, the power dissipated in the resistor:
   a) Increases
   b) Decreases
   c) Remains constant
   d) None of the above
   Answer: option (b).
5) Consider an electric circuit with a voltage source V connected to a resistor R with ideal conducting wires (zero resistance). Which of the following is true?
   a) Most of the energy transfer from the voltage source to the resistor occurs through the ideal conducting wires
   b) Most of the energy is transported to the resistor parallel to and just outside the conducting wires
   c) Energy is not transferred from the voltage source to the resistor
   d) The current given by the voltage source is consumed in the resistor with the help of the conducting wires
   Answer: option (b).

Problems

1) Calculate the magnitude of the drift velocity of the electrons when 5 A of current exists in a material with a 2 mm^2 cross-section. The number of free electrons in 1 cm^2 of that material is 8.5×10^{22}.

Solution:

We know that the current density is given by

$$|\vec{J}| = ne\,|\vec{v}_d|$$

$$\Rightarrow |\vec{v}_d| = \frac{|\vec{J}\,|}{ne} = \frac{i}{neA}.$$

Upon substituting the values given in the problem, we obtain that $|\vec{v}_d| = 0.18\ \text{mm s}^{-1}$.

2) Calculate the resistance of an aluminum wire of length 100 cm and cross-sectional area 2.0 mm^2. The resistivity of aluminum is $2.6 \times 10^{-8}\ \Omega$ m.

Solution:

The resistance is $R = \rho\frac{1}{A}$. Hence, we have

$$R = \frac{2.6 \times 10^{-8} \times 1}{2 \times 10^{-6}} = 0.013\ \Omega.$$

3) Find the equivalent resistance between points A and B in figure M5.1.

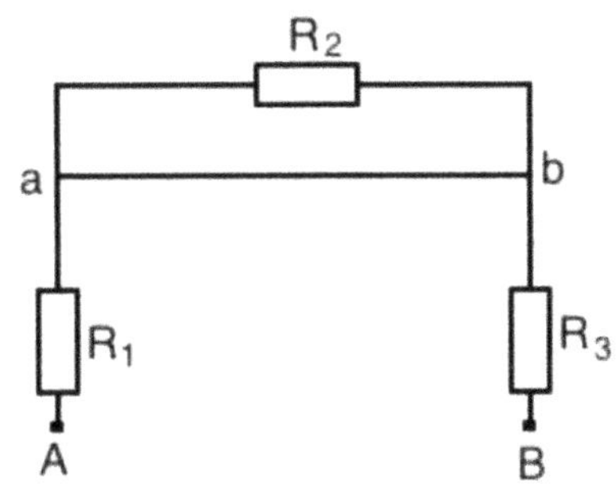

Figure M5.1.

Solution:

Let us assume that the resistance of the connecting wire is r. The connecting wire ab and the resistor R_2 are in parallel. The equivalent resistor between the nodes (a) and (b) is then given by

$$R_{ab} = \frac{rR_2}{r + R_2}.$$

We now let the resistor r tend to zero since it is an ideal wire. We thus have for the ideal case

$$R_{ab} \lim_{r \to 0} \frac{rR_2}{r + R_2} = 0.$$

Hence, the resistance between the points (a) and (b) is zero. Thus, the resistance between points A and B is the series combination of R_1 and R_3 and we have

$$R_{AB} = R_1 + R_3.$$

4) Find the equivalent resistance between the points A and B shown in M5.2.

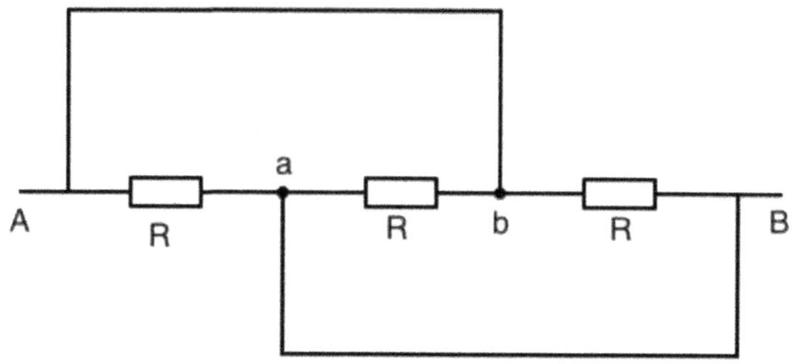

Figure M5.2.

Solution:

It can be seen from the figure that points *A* and (*a*) are at the same potential. And points *B* and (*b*) are at the same potential. Hence, the above circuit can be redrawn as shown in figure M5.2A.

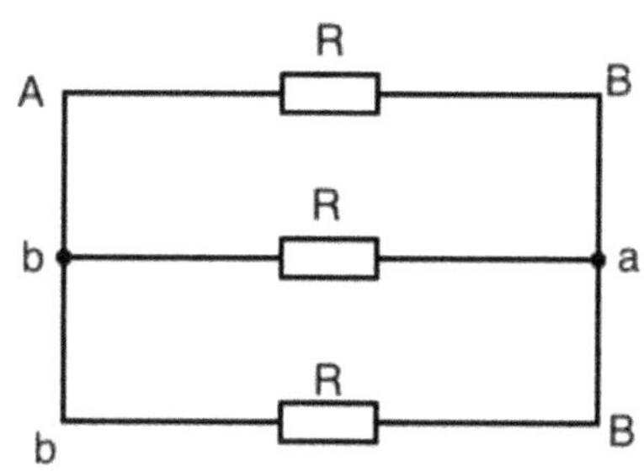

Figure M5.3.

It can easily be seen that the equivalent resistance between the points *A* and *B* is *R*/3.

5) Find the charge on one of the plates of the parallel plate capacitor *C* in the circuit shown in figure M5.3.

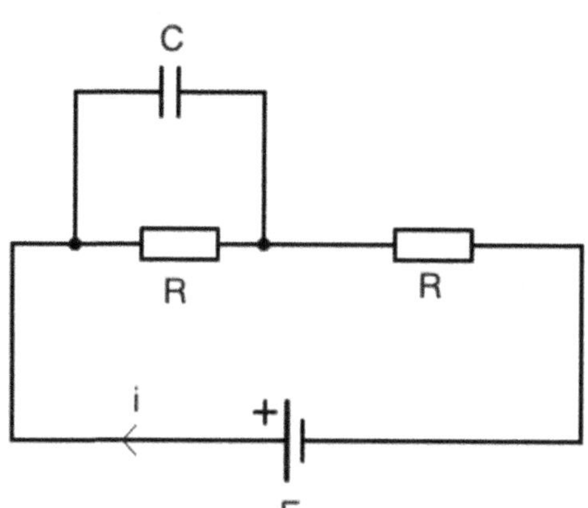

Figure M5.3.

Solution:

In the steady state, the capacitor acts as an open circuit to DC current. Thus the current through from the battery E is

$$i = \frac{E}{2R}.$$

The voltage V_1 across the first resistor then is given by

$$V_1 = \frac{E}{2R}R = \frac{E}{2}.$$

This is also the potential difference between the plates of the capacitor. Hence, the charge Q in one of the plates of the capacitor is given by

$$Q = CV_1 = \frac{EC}{2}.$$

Chapter 6

MCQs:

1) A charged particle moves in a gravity-free space without change in velocity in an electric field E and a magnetic field B. Which of the following is not possible?
   a) $E = 0,\ B = 0$
   b) $E = 0,\ B \neq 0$
   c) $E \neq 0,\ B = 0$
   d) $E \neq 0,\ B \neq 0$
   Answer: option (c)
2) Consider the situation shown in figure M6.1. The straight current-carrying wire closed at infinity is fixed, while the loop can move freely. The loop will:

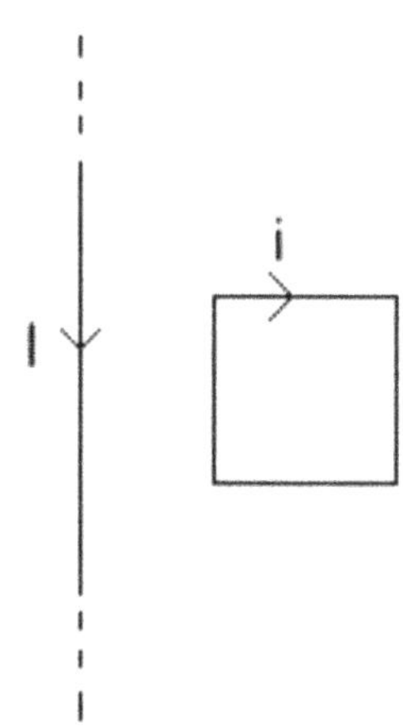

Figure M6.1.

a) Remain stationary
b) Move toward the wire
c) Move away from the wire
d) None of the above
Answer: option (c).

3) A long straight wire carries a current along the Z-axis. One cannot find two points in the XY plane such that:
a) The directions of the magnetic fields are the same
b) The magnetic fields are equal
c) The field at one point is opposite to that at the other point
d) The magnitudes of the magnetic fields are equal
Answer: option (b).

4) A vertical wire carries a current in the upward direction along the positive Z-axis. A proton beam sent horizontally towards the wire will be deflected:
a) Towards the negative Z-axis
b) Towards the positive Z-axis
c) Towards the positive Y-axis
d) None of the above
Answer: option (a).

5) Two infinitely long current-carrying wires carrying currents i_1 and i_2 are placed parallel to each other. Suppose that when the currents are in the same direction, the magnetic field created midway between the wires is k_1. When the currents are in opposite directions, suppose that the magnetic field created at the midpoint is k_2. The ratio i_1/i_2 is equal to:
a) k_1/k_2
b) k_2/k_1
c) $(k_1 + k_2)/(k_2 - k_1)$
d) $(k_1 + k_2)/k_1$
Answer: option (c).

Problems

1) A charge of 2 nC moves with a speed of 200 m s^{-1} along the positive X-axis. A magnetic field $\vec{B} = (\hat{j} + \hat{k})T$ exists in space. Find the magnetic force on the charge.

Solution:

The magnetic force is given by $\vec{F} = q(\vec{v} \times \vec{B})$. Upon the substitution of values, we have that

$$\vec{F} = 2 \times 10^{-9} \times \left(200\hat{\imath} \times \left(\hat{j} + \hat{k}\right)\right)$$

$$\Rightarrow \vec{F} = 400 \times 10^{-9} \times \left(\hat{\imath} \times \hat{j} + \hat{\imath} \times \hat{k}\right)$$

$$\Rightarrow \vec{F} = 400 \times 10^{-9} \times \left(\hat{k} - \hat{j}\right) \text{N}.$$

2) A long solenoid is formed by winding 40 turns cm s^{-1}. What current is necessary to produce a magnetic field of 10 mT inside the solenoid?

Solution:

The magnetic field inside a long solenoid is given by $B = \mu_0 ni$. Upon the substitution of values, we find that $i = 2$ A.

3) Two long straight wires, each carrying a current of 10 A are kept parallel to each other at a distance of 5 cm. Find the magnitude of the magnetic force experienced by 10 cm of a wire.

Solution:

The magnitude of the magnetic field at a wire due to another wire is given by

$$B = \frac{\mu_0 i}{2\pi r}.$$

Upon the substitution of values, we find that $B = 4 \times 10^{-5}\, T$. The magnitude of the force experienced by 10 cm of this wire due to the other is

$$F = ilB = 4 \times 10^{-5}\,\text{N}.$$

4) Suppose a magnetic field B exists in space such that it is directed into the plane of this paper. A particle of mass m is projected with a speed v in the plane of paper. The velocity is perpendicular to the magnetic field. Find the revolution frequency of the particle.

Solution:

The force F is perpendicular to both v and B. The particle moves in circles with a radius r. The magnetic force provides the necessary centripetal force for this circular motion. Thus, we have for circular motion that

$$qvB = m\frac{v^2}{r}$$

$$\Rightarrow r = \frac{mv}{qB}.$$

The time taken to complete the circle is $T = \frac{2\pi r}{v}$. Thus, we have

$$T = \frac{2\pi m}{qB}.$$

The frequency of revolution is $f = 1/T$. Thus,

$$f = \frac{qB}{2\pi m}.$$

It is interesting to note that this does not depend on the speed v. This is called the cyclotron frequency.

5) A proton with charge e moves with a constant speed v along a circular path of radius r. Find the magnetic moment of the circulating proton.

Solution:

First we need to find out the equivalent current due to the motion of the proton in the circular path. Consider a point on the circle. The proton crosses the circle $v/2\pi r$ number of times per unit time. Hence the electric current is

$$i = \frac{ev}{2\pi r}.$$

The area enclosed by this circular loop is πr^2. Hence the magnetic moment is given by

$$\mu = iA = \left(\pi r^2\right)\frac{ev}{2\pi r} = \frac{evr}{2}.$$

Chapter 7

MCQs:

1) A bar magnet is released from rest along the axis of an infinitely long vertical copper tube. After sufficient time the magnet will:
   a) Stop
   b) Move with almost constant velocity
   c) Accelerate downward continuously
   d) Oscillate
   Answer: option (b).
2) Consider the circuit shown in figure M7.1. A conducting loop ABCD is placed next to the circuit. When the switch S has been closed and then after some time opened again, the closed loop will show:

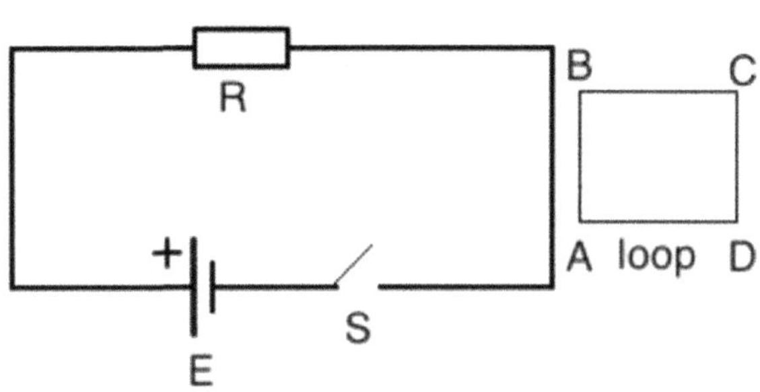

Figure M7.1.

   a) A clockwise current pulse
   b) An anti-clockwise current pulse

c) A clockwise current pulse and an anti-clockwise current pulse
d) An anti-clockwise current pulse and a clockwise current pulse.
Answer: option (c).

3) A conducting loop is placed in a magnetic field B. Which of the following processes does not induce an emf in the loop?
   a) The loop is translated from one point in space to another
   b) The loop is rotated about a diameter
   c) The loop is deformed
   d) The loop is stretched and its area increases
   Answer: option (a).
4) Moving a conductor in a constant magnetic field can induce an emf in the conductor. Is this statement true or false?
   Answer: True.
5) The magnetic field energy in an inductor changes from the maximum value to the minimum value in 5 ms. The frequency of the AC source is:
   a) 20 Hz
   b) 30 Hz
   c) 40 Hz
   d) 50 Hz

Problems

1) A rectangular loop ABCD is placed in a magnetic field B such that the side AB is in the field and the side CD has a resistor or resistance R. The field is parallel to the axis of the loop. A force F is applied to the loop and the loop is pulled out with a uniform speed v. Find the power dissipated in the resistor.

   Solution:

   The emf induced in the loop is due to the motion of the wire AB. The other three sides do not contribute to the production of emf in the loop. Hence, the current in the loop is given by

   $$i = \frac{E}{R} = \frac{vBl}{R}.$$

   The power dissipated is given by $P = i^2R = \frac{v^2B^2l^2}{R}$.
2) An average induced emf of 1 V appears in a coil when the current in it is changed from 1 A in one direction to 1 A in the opposite direction in 1 s. Find the self-inductance of the coil.

   Solution:

   The average rate of change of the current is $\frac{\mathrm{d}i}{\mathrm{d}t} = \frac{1-(-1)}{1} = 2\ \mathrm{A\ s^{-1}}$
   We know that $|E| = L\frac{\mathrm{d}i}{\mathrm{d}t}$. Hence, we obtain that $L = 0.5$ H.
3) Calculate the energy stored in an inductor of inductance 100 mH when a current of 1 A is passed through it.

Solution:

The energy stored is given by

$$U = \frac{1}{2}Li^2$$

$$\Rightarrow U = 0.05 \text{ J}.$$

4) A conducting circular loop is placed in a uniform magnetic field $B = 0.01$ T with its plane perpendicular to the field. Let us suppose now that the radius of the loop starts increasing at a rate of 1 mm s^{-1}. Find the induced emf in the loop at an instant when the radius is 1 cm.

Solution:

Let the radius be r at a time t. The flux of the magnetic field at this instant is

$$\Phi = \pi r^2 B.$$

We then have that

$$\frac{\mathrm{d}\Phi}{\mathrm{d}t} = 2\pi r B\frac{\mathrm{d}r}{\mathrm{d}t}.$$

The induced emf is therefore given by

$$E = \frac{\mathrm{d}\Phi}{\mathrm{d}t} = 0.625 \ \mu\text{V}.$$

5) Consider a circuit in which an inductance L and a resistor R are connected to a battery of emf E. Find the maximum rate at which energy is stored in the magnetic field.

Solution:

The energy stored in the magnetic field is

$$U = \frac{1}{2}Li^2 = \frac{1}{2}Li_0^2\left(1 - \mathrm{e}^{-t/\tau}\right)^2.$$

The rate at which energy is stored in the inductor is

$$\frac{\mathrm{d}U}{\mathrm{d}t} = \frac{Li_0^2}{\tau}\left(\mathrm{e}^{-t/\tau} - \mathrm{e}^{-2t/\tau}\right).$$

For the rate to be maximum we need that

$$\frac{\mathrm{d}^2U}{\mathrm{d}t^2} = 0$$

$$\Rightarrow \mathrm{e}^{-t/\tau} = \frac{1}{2}.$$

Thus, we have $\left.\frac{\mathrm{d}U}{\mathrm{d}t}\right|_{\max} = \frac{Li_0^2}{\tau}\left(\frac{1}{2} - \frac{1}{4}\right) = \frac{E^2}{4R}.$

Chapter 8

MCQs:

1) In the Coulomb gauge, the magnetic field can be produced by:
   a) Moving charge
   b) A changing electric field
   c) Both
   d) None of the above
   Answer: option (c).
2) In the Lorentz gauge, the magnetic field is produced by:
   a) Moving charge
   b) A changing electric field
   c) Both
   d) None of the above
   Answer: option (a).
3) Displacement current goes through the plates of the capacitor and equals the current when the capacitor is being:
   a) Charged
   b) Discharged
   c) Both
   d) None of the above
   Answer: option (c).
4) If μ_0 represents the permeability and ε_0 represents the permittivity of free space, then the speed c of an electromagnetic wave in vacuum is given by:
   a) $c = \sqrt{\varepsilon_0 \mu_0}$
   b) $c = \dfrac{1}{\sqrt{\varepsilon_0 \mu_0}}$
   c) $c = \dfrac{1}{\varepsilon_0 \mu_0}$
   d) $c = \varepsilon_0 \mu_0$
   Answer: option (b).
5) Which of the following produces electromagnetic waves?
   a) A stationary charged particle
   b) A uniformly moving charged particle
   c) An accelerated charged particle
   d) None of the above
   Answer: option (c).

Problems

1) Calculate the energy $\mathrm{d}W$ passing during a time $\mathrm{d}t$ through a unit area perpendicular to the direction of the propagation of an electromagnetic wave.

   Solution:

   The quantity $\mathrm{d}W$ can be represented with the help of Poynting vector as

$$dW = S\ dt,$$

This implies that it can be written as

$$dW = \frac{EB}{\mu_0}\ dt.$$

The instantaneous values of E and B are connected by the following relationship

$$E = cB.$$

Thus, we have $dW = \sqrt{\frac{\varepsilon_0}{\mu_0}}E^2\ dt$.

2) The maximum electric field in a plane electromagnetic wave is 100 N/C. The wave is going in the x-direction and the electric field is in the y-direction. Find the maximum magnetic field in the wave and its direction.

Solution:

We have seen before that the instantaneous values of the electric and magnetic field are related by the equation

$$E = cB.$$

Thus, both magnetic and electric fields achieve their maximum simultaneously. Thus, we have that

$$E_0 = cB_0.$$

Upon the substitution of values, we find that $B_0 = 0.33 \times 10^{-6}$ T.

3) Suppose that a parallel plate capacitor with a plate area A and a distance d between the plates is charged by passing a constant current I. Consider the surface S of the area A/k parallel to the plates and drawn symmetrically between them. Find the displacement current through the area S.

Solution:

Let us assume that the charge on the capacitor at time t is Q. The electric field between the plates is then given by

$$E = \frac{Q}{\varepsilon_0 A}.$$

The flux through the surface is then given by

$$\Phi = E\frac{A}{k} = \frac{Q}{k\varepsilon_0}.$$

The displacement current then is given by

$$i_d = \varepsilon_0\frac{d\Phi}{dt} = \frac{\varepsilon_0}{k\varepsilon_0}\frac{dQ}{dt} = \frac{i}{k}.$$

4) Protons with the same velocity v in the positive x-direction form a beam with current I. Find the magnitude and direction of the Poynting vector **S** outside the beat and at a distance d from the beam.

Solution:

The electric field is directed perpendicularly to the beam and the magnetic field is given by the right-hand thumb rule. It can be seen that the direction of the Poynting vector is same as the direction of the velocity vector **v**.

We know that the electric field of an infinitely charged line of linear charge density λ is given by

$$E = \frac{\lambda}{2\pi d\varepsilon_0}.$$

We can use Ampere's law to determine the magnetic field at a distance d from the beam. We obtain

$$B = \frac{I\mu_0}{2\pi d}.$$

Hence the Poynting vector $S = EB/\mu_0$ is given by

$$S = \frac{1}{\mu_0}\frac{I\mu_0}{2\pi d} \times \frac{\lambda}{2\pi d\varepsilon_0}.$$

Upon substitution of the fact that $I = \lambda v$, we obtain that $S = \frac{I^2}{4\pi^2 d^2 \varepsilon_0 v}$.

5) The energy crossing per unit area per unit time perpendicular to the direction of propagation is called the intensity of an electromagnetic wave. Derive an expression for the intensity of the electromagnetic wave if the electric and magnetic fields in the wave are described as

$$E = E_0 \sin\omega\left(t - \frac{x}{c}\right)$$

$$B = B_0 \sin\omega\left(t - \frac{x}{c}\right).$$

Solution:

The energy of the electric field in a small volume $\mathrm{d}V$ is given by

$$U_E = \frac{1}{2}\varepsilon_0 E^2 \,\mathrm{d}V.$$

Similarly, the energy of the magnetic field in the volume $\mathrm{d}V$ is given by

$$U_B = \frac{1}{2\mu_0} B^2 \,\mathrm{d}V.$$

Thus, the total energy per unit volume is given by

$$u_T = \frac{1}{2}\varepsilon_0 E^2 + \frac{1}{2\mu_0}B^2.$$

We now substitute the field equations in the above formula and take the average of $\sin^2$ to be equal to ½ to obtain the average value of the total energy to be

$$u_{\text{avg}} = \frac{1}{4}\varepsilon_0 E_0^2 + \frac{1}{4\mu_0}B_0^2.$$

Using the equations $E_0 = cB_0$ and $c = \dfrac{1}{\sqrt{\varepsilon_0\mu_0}}$, we obtain that

$$u_{\text{avg}} = \frac{1}{2}\varepsilon_0 E_0^2$$

The energy intensity is given by $I = u_{\text{avg}}c$. Hence, we obtain that

$$I = \frac{1}{2}\varepsilon_0 E_0^2 c$$

Chapter 9

MCQs

1) The standard derivation of Kirchhoff's voltage law assumes that the ___________ law of thermodynamics is satisfied in each loop.
   a) First
   b) Second
   c) Zeroth
   d) None of the above
   Answer: option (a).
2) The work done by an external agent in moving a positive test charge around a closed loop of a circuit against ___________ electric field/s is zero.
   a) Non-conservative
   b) Conservative
   c) The superposition of conservative and non-conservative
   d) None
   Answer: option (b).
3) Which of the following quantities do not change when a resistor connected to a battery is heated due to the current?
   a) Drift speed of the electrons
   b) Resistivity
   c) Resistance
   d) None
   Answer: option (d).
4) The conversion of electrical energy into Joule heat in a resistor is:
   a) A reversible process
   b) An irreversible process

c) Both
d) None of the above
Answer: option (b).

5) Assuming that only KCL is true, the stationary distribution of currents in a linear electric circuit is governed by the principle of ______________.
   a) Maximum entropy
   b) Maximum entropy production
   c) Minimum entropy
   d) The second law of thermodynamics
   Answer: option (b).

Problems

1) A wire of resistance 100 Ω is bent to make a complete circle. Find the resistance between two diametrically opposite points.

   Solution:

   We have that the resistance of a wire is given by

$$R = \rho \frac{l}{A}.$$

   Between two diametrically opposite points AB, the length is reduced to half while the other parameters remain constant. Hence, there are two resistors connected in series with a resistance of 50 Ω each. Thus, the equivalent resistance of diametrically opposite points is 25 Ω.

2) Find the current through the resistor R_2 in the circuit shown in figure M9.1 using the superposition theorem. Also find out which voltage source supplies current to resistor R_2.

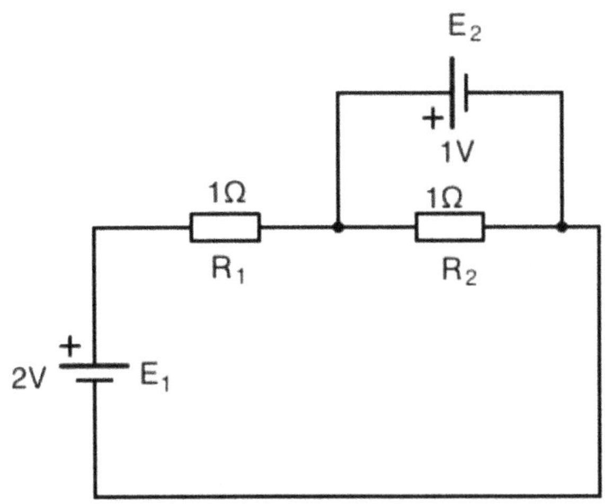

Figure M9.1.

Solution:

• Shorting the voltage source E_1

The current from the voltage source E_2 is 2 A and 1 A goes through resistor R_2 and 1 A through resistor R_1.

• Shorting the voltage source E_2

The current from the voltage source E_1 is 2 A and all of the current in this case goes through the short-circuited voltage source E_2.

Hence, the current through resistor R_2 is 1 A and it is supplied by voltage source E_1.

3) Consider the circuit shown in figure M9.2. A current of 1 A flows through resistor R. Consider two points a and b on the ideal wire. The potential difference between a and b is zero as the wire is ideal and has no resistance. How can the current flow between the two points a and b if the potential difference is zero? Justify your answer.

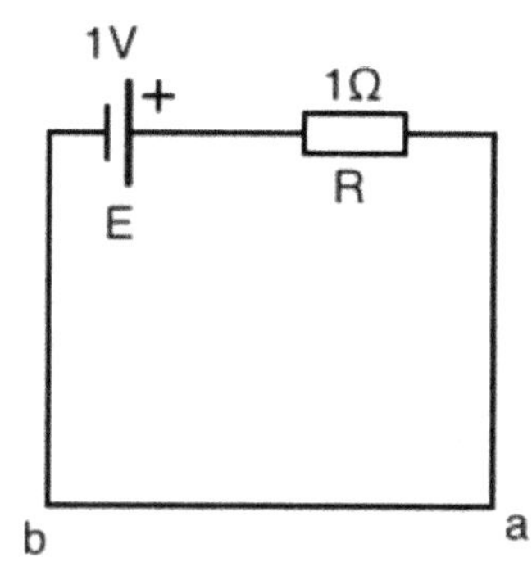

Figure M9.2.

Solution:

The question can be answered easily if we take the surface charges into account. The surface charges are arranged in the wire between (a) and (b) such that there is no component of electric field parallel to the wire. This is because the resistance of the wire is zero and hence according to Ohm's law there should be no component of electric field in the direction of the current density. The non-conservative forces inside the battery E force the charges to leave the battery and the electrons move with constant velocity because of their inertial motion. In resistor R, the electric field created by the surface charges allows the electrons to move in the same drift velocity. Hence, the current flows through the entire circuit.

4) Consider the circuit shown in figure M9.3. Find the product of the ambient temperature of the resistors and the rate of the entropy produced in the circuit.

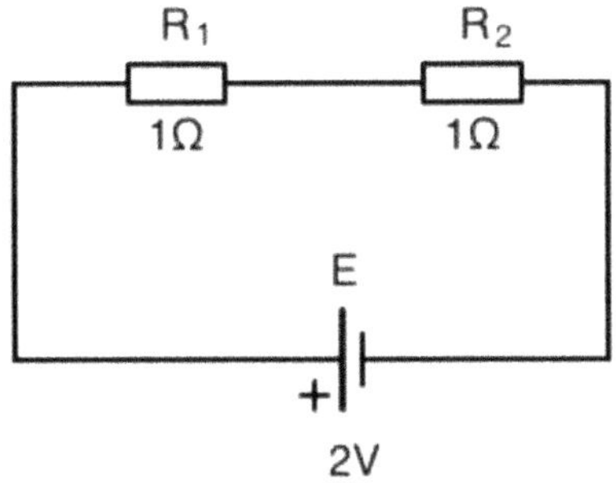

Figure M9.3.

Solution:

We have already seen that the entropy production times the ambient temperature is just equal to the Joule heat power:

$$T_{\text{amb}}\frac{\mathrm{d}S}{\mathrm{d}t} = P.$$

The current i that flows through resistors R_1 and R_2 is 1 A. Since entropy is an additive quantity, we obtain that

$$T_{\text{amb}}\frac{\mathrm{d}S}{\mathrm{d}t} = i^2R_1 + i^2R_2 = 2 \text{ W}.$$

5) For a circuit consisting of only a single loop, prove that the validity of Tellegen's theorem reduces to the validity of Ohm's law.

Solution:

For a single loop, we can write the equivalent resistance and equivalent voltage of any number of combinations of resistors and voltage sources as R_{eq} and E_{eq}, respectively. Now, Tellegen's theorem states that the sum of the product of voltages across and current through different branches is zero. Hence, we have for our circuit that

$$\sum vi = -E_{\text{eq}}i + i^2R_{\text{eq}} = 0$$

$$\Rightarrow E_{\text{eq}} = iR_{\text{eq}}.$$

This is just Ohm's law. Hence, we have proved that for a single loop the validity of Tellegen's theorem just reduces to the validity of the Ohm's law.

Lightning Source UK Ltd.
Milton Keynes UK
UKOW07n0159060517
300580UK00004B/42/P

9 780750 312677